St. George Jackson Mivart

The Helpful Science

St. George Jackson Mivart

The Helpful Science

ISBN/EAN: 9783337035686

Printed in Europe, USA, Canada, Australia, Japan

Cover: Foto ©berggeist007 / pixelio.de

More available books at **www.hansebooks.com**

THE HELPFUL SCIENCE

BY

ST. GEORGE MIVART, F.R.S.

NEW YORK
HARPER & BROTHERS PUBLISHERS
1895

THE HELPFUL SCIENCE

Part I

It has been said reproachfully that man is fond of novelty. The reproach is unreasonable, since change is essential to both our bodily and mental health. Only through changes in its surroundings is the consciousness of the child first awakened, and only by the aid of fresh external or internal modifications is consciousness maintained in activity.

Moreover, though analogous social and political conditions frequently recur, it is impossible that any past experience can ever be truly repeated, and this not only on account of the ceaseless changes of our environment, but also from the very conditions of our bodily frame. A nerve continuously stimulated fails, after a time, to respond to the stimulus, and thus it is that our appetite becomes jaded by an unceasing supply of what might at first have been the object of keen pursuit, and positive aversion often succeeds desire. Our taste changes again and again as we travel from childhood to old age, and education and culture notoriously modify man's inclinations and sentiments.

Thus it is simply inevitable that the feelings, tastes, and intellectual occupations of succeeding centuries must always be more or less divergent, although analogous recurrences may again and again take place.

That conditions should so recur is at least as desirable as it is inevitable. For of the many objects which merit our attention, all cannot at the same time be attended to in the degree they merit. Thus it is that in each age some objects may be not only unduly neglected, but even disliked, so that they need to be subsequently recalled to the attention of succeeding times. After the æsthetic outburst of Hellas came the wondrous jurisprudence and political sagacity of Rome, paving the way for those noble creations of mediævalism, feudal honor, devoted asceticism, majestic worship, and the keenest polish which the human intellect has ever received.

Then blossomed that great "recurrence," the Renascence, once more bringing Plato into fashion, with a love for physical science akin to that which had existed in the mind of Aristotle. Therewith there naturally arose a feeling of aversion from what had charmed men before, with a neglect of one department of nature, just as another department had been previously neglected by mediæval students.

Political freedom and social amelioration were the next objects of pursuit among civilized mankind, and they have continued, with physical science, to be the main occupation of those not absorbed in seeking wealth; for wealth, as a source of pleasure and for its own sake, has doubtless been the pursuit of most men since the first human society arose.

The occupations of the immense majority of that part of mankind which is raised above the sordid need of seeking its daily bread, and is not merely devoted to pleasure, may be said to be travel, politics, philanthropy, the promotion of progress, useful knowledge, physical science, and art — all, variously and in different degrees, good and laudable. They are also practical pursuits. They are concerned with "doing" as well as with "knowing." What is *practical* has special charms for us, for the English-speaking races of mankind are, in all climes, eminently *doers*. Our profound respect for work, for "doing" (as has been before remarked), is shown by the familiar usage of our mother-tongue. We English speakers greet each other with, "How do you *do?*" A Frenchman says, with an implied regard for appearances, "How do you *carry yourself?*" A German ascends at once to the highest abstractions, and asks, "How goes *it?*" A Spaniard, as one who first crossed the Atlantic, inquires, "How do you *go?*" while the Italian, for so long a time unprogressive, says, "How do you *stand?*" or even insinuates the *dolce far niente* by asking, "How do you *exist?*" Only the English-speaking man seems instinctively to feel that you cannot be in a satisfactory state unless you are *doing* something. To those who are aware they so feel, and are therewith content, these papers are addressed. Their aim is to be eminently practical.

But readers who know anything of the present writer may reply: "You are not practical yourself. You are no politician or mechanician, no lawyer or medical practitioner, not even an artist; you pursue science for its own sake!" The statement is true, but has no force as an objection; for we know how often the most solid and extensive practical gains have been due to abstruse and seemingly most unpractical exertions of thought and endeavor. In America useful deductions from abstruse studies have been exceptionally developed, and the whole civilized world is being lighted—and men are invited to read, to play, to pray, or to sin— through the help of Franklin's brain.

This will be universally assented to, for the value and the trustworthy progress of physical science are unquestioned. Many foolish discussions are carried on in the world about us; but certainly no one is so foolish as to question the value of such science or the fact of its progress. Certainly I, who have loved it from my earliest years and devoted such small powers as I possess to its service, have no disposition to undervalue it. I well know that the study of living organisms has, since I can recollect, made great progress, and I see grounds for absolute certainty now about many zoölogical facts and laws which were doubtful or undreamed of when I was a lad.

Nevertheless such science is not everything, and I am profoundly convinced that there are deeper questions which merit the best attention of all practical men.

Some may here exclaim: "Ah! you mean religion. It is true that, with very many in Europe, it has gone entirely out of fashion, save as a ray of emotion gilding some philanthropic sentiment; but it is not so in America, where so many creeds count such multitudes of followers, and where respectable citizens have, as a rule, respectable church-sittings."

But my object is neither to champion nor to combat any special creed, but to strictly confine myself to the sufficiently wide domain of science external to denominational theology, though it may be we shall find that pure reason affords a sure basis for the fundamental beliefs of all theology. I mean, therefore, nothing to which any section of my readers can, I venture to think, reasonably object, whatever their theological beliefs or disbeliefs may be. I desire to eliminate all such questions and to confine myself purely and simply to matters which regard well-ascertained scientific truths. I desire but to call attention to propositions which must, I believe, be assented to by every consistent lover of science who is convinced that at least some scientific truths have been brought to our knowledge—truths on which we can with entire confidence rely. But what I refer to has a much wider scope than even physical science. It concerns religious devotees no less than followers of fashion or men who seek their many-million pile. It concerns no creed, indeed, directly, nor any pleasure, policy, or art; and yet it alone gives value to each and all of them. The essence of a creed is something to believe; yet how can we believe what we in no way know?

It may be objected that "belief may exceed knowledge." Yet no one out of Bedlam can believe anything save on some grounds; and he ought to know the grounds of his belief, and be able to give an account of the faith that is in him. Our science, then, concerns no particular creed, but all creeds; no art, but all arts; no science, but all sciences; no rule of conduct, but all rules of conduct. It is the most fundamental, but also the most far-reaching; the most central, yet the most centrifugal; the most intensely personal and, nevertheless, the most universal of all questions and of all truths.

As it must underlie all our pleasures no less than all our knowledge, it might be called, what indeed it has been termed, "the Gay Science." Yet with our love for what is practical, and with our recognition that we have to be "up and doing," I prefer—since it supplies the only solid ground for our activity—to call it *The Helpful Science.*

But since so much good and solid work has been and is being done in all departments of science and art; since social amelioration goes on apace and philanthropy was never more active; since, also, no good work can be done without knowing how to do it, it would seem to follow that this "helpful science" must be already so widely known that to write about it (as here proposed), can but be superfluous. Several meanings, however, attach to the expression "to know," and we may and do know many things without our ever having explicitly adverted to and recognized the fact that we do know them.

Socrates, as we have all read, used, when addressing the men of Athens, to call himself a "midwife;" by which he meant that his business was not really to teach them, but to draw forth from and exhibit to his hearers truths which they bore unconsciously in their own bosoms. I would humbly follow in the footsteps of that great man, and endeavor to explicitly set forth certain very fundamental verities which we all possess implicitly, though we may never have had the eyes of our intellect directly turned upon them.

But here I may at once be met by this objection and a protest: "You say you wish to imitate Socrates, but he was a philosopher. Now we are quite willing to be talked to about science and art, politics, morals, or philanthropy; but, if you please, no 'philosophy' for us! We detest metaphysics and will have nothing to do with your 'quiddities' and 'essences,' 'the transcendental' and 'the absolute,' all which we consign with a hearty execration to the bottom of the Dead Sea!"

I, nevertheless, entreat my readers to grant me a modicum of tolerance and to read on a little more. I can ask this with a better grace since I deeply sympathize with the feeling which might dictate such a protest as that I have just supposed.

It is indeed by no means wonderful that men should turn away in disgust from the mere sight or sound of the word "metaphysics"—above all, that English-speaking men and women should do so. An eminently practical set of people, however energetic, could hardly be disposed to work, whether with hand or brain, without a fair promise of some good result—like a squirrel in a turning cage. And certainly no much better result can be expected from spending time over "philosophy," as it is set before us in most modern works on that subject. Philosophy, it must also be confessed, has for the last few centuries had little progress to boast of. Moreover it has been presented in a very repellent manner. Of course every branch of knowledge has, and must have, its own technical terms, but it would seem as if the very spirit of evil-speaking had taken possession of most metaphysicians, and as if they had deliberately sought to make themselves as unintelligible as possible to the ordinary run of mankind.

Every science and each art has, as before

said, its technical terms. The anatomist and the lawyer respectively find it easier to speak of the "fifth ventricle" or the "Cestui que trust," than to denote such conceptions by descriptive phrases in plain words. Nevertheless, with a little circumlocution most scientific statements could be made in ordinary language, and I believe that all philosophical conceptions can be so expressed, if only a little trouble is taken by writers on the subject. At all events the present writer thinks he can promise to at least make himself intelligible, and to refrain from the use of bewildering words in treating of a matter which readers will in the end; it is believed, see to be, as before said, a very practical one.

But will the Socratic method really suffice? We cannot extract sunbeams from cucumbers! Can we draw forth philosophy from clod-hoppers?

Some men—we will even dare to say some women—are foolish, while others are exceptionally gifted; but plain, ordinary reason is possessed by the overwhelming majority of men and women. Such ordinary reason, with a little patience and perseverance, are all that is needed for a good insight into "the helpful science."

For just as science is but ordinary reason carefully and exactly applied to the examination of what surrounds us, so philosophy is nothing but the same reason applied to what is most fundamental in all the other sciences—on which account it may be called "the science of sciences."

No educated man should rest satisfied without trying to understand it, for otherwise his knowledge, sufficient as it may be for many purposes, must none the less be without a solid, logical basis. There are certain questions which underlie all physical science, and we are persuaded that not a few of the readers of this volume will like to see drawn out what those questions are, and what are, therefore, the necessary foundations of all our knowledge. I have spoken of "educated men," but my experience convinces me that a man may be a good philosopher with very little education. In considering the vulgar we are very apt to be misled by small matters which may strongly impress the imagination, but which our judgment, on reflection, must own to be trivial. We are sometimes tempted to despise an intellect which manifests itself only by uncouth gestures and coarse speech, wherein the rules both of correct pronunciation and grammar are violated; and yet that intellect may be, in fact, quite as good as our own. I have more than once been surprised, when talking with peasants, to find how correct was their appreciation even of questions of philosophy, when once the difficulties arising simply from our different modes of expressing essentially similar ideas had been surmounted. If this is the case with farm-laborers, there must be plenty of unconscious philosophers among the keen-witted artisans of our large cities who take an interest in deep questions. If, therefore, every such man chooses to upbraid me with being a "philosopher," I can without scruple respond by a *Tu quoque.* For, in fact, every sane man must, consciously or unconsciously, have some system of philosophy whether he will or no. But a reasonable man will try and get as reasonable a view of things as he can, and to such a man it will seem worth while to spend a little time and take a little trouble with respect to questions which constitute the foundation of all his knowledge.

The present writer feels a great satisfaction in addressing an American public on such questions, because he not only knows that education is highly prized and widely diffused in the United States, but he also believes that in a nation where every form of knowledge is eagerly sought after there must be a vigorous appetite also for philosophy as soon as its real nature is understood. In America there seems to be a strong desire to get to the bottom of matters and to know what things are, even though the *how* and the *wherefore* of their being what they are may remain inscrutable. There is a healthy appetite for *facts*, with a remarkable absence of prejudice and a hearty wish to be "thorough."

But the helpful science does not alone concern "knowing," but "doing" also. It is at the foundation of ethics and therefore of conduct, and thus considered its claim on the attention of the English-speaking races is exceptionally great. For they have arrived at a degree of self-government, accompanied by a combination of order and freedom, which has hitherto been unknown elsewhere. In our race, then, it must evidently be a matter of the most extreme importance that opinion should not be misled in matters which touch so nearly the innermost springs of morality and social life. It is supremely important that there should be no doubt as to the validity of those declarations of reason upon which our moral perceptions and all our social regulations ultimately depend. It may be thought that metaphysical questions concern the highly educated alone. But this opinion is a very mistaken one, for metaphysical doctrines (it may be in crude forms) soon filter down to, and have their full effect on, the lowest classes of society. In some form or other they come home to almost every one.

There are few men indeed who have never

thought about either their origin or their destiny, or the wonderful mystery of life. In hours of sadness and trial, when some object, aimed at for years, is suddenly seen to be hopelessly beyond reach; when the symptoms of a fatal disease first become clearly recognized, or when we stand and gaze on the marble features of our much-loved dead, deep questions not seldom arise in the sorrow-laden mind: What and whence are we? Why are we here? What should be our true end? What is the real meaning, what the good of life?

Of course there are multitudes of men who, save at such solemn moments, put such questions on one side, and get into a habit of going through life's daily tasks, taking what pleasure they may on the road and as much of it as they can get. But, at least, the men and women who address themselves to the perusal of an essay such as this are not likely to be anxious to shirk all serious questions, or indisposed to take a little trouble if they see that such labor need not be fruitless; while they will know well that no good work can be done without taking some trouble about it.

The task here undertaken is to show that the helpful science of philosophy is not only a basis and support absolutely necessary for every physical science, but that it is no less, as before said, the foundation of ethics and of all right conduct.

Nevertheless no appeals will here be made to sentiments and feelings, whether moral or otherwise. Appeal will be made to the pure intellect of readers and to nothing else. And, indeed, I desire to take this opportunity plainly to declare that not only here and now, but everywhere and always, I unhesitatingly affirm that no system can or should stand which is unable to justify itself to reason. I possess no faculty myself, nor do I believe that any human faculty exists, superior to the intellect, or which has any claim to limit or dominate the intellect's activity. Feelings and sentiments have their undoubted charm and due place in human life, but that place is a subordinate one, and should be under the control of right reason.

Yet it is by no means only against those who would undervalue reason for the sake of sentiment that these lines are written; their object is to uphold the rights of our rational nature against all who, from whatever side or in the name of whatsoever authority, would either impugn its sovereign claim upon our reverence or would unduly restrict the area of its sway.

Five hundred years ago the study of philosophy was the intellectual fashion of the day, and, as we before said, never were the wits of mankind so well polished, never was the intellect provided with a keener razor's edge — to use the best material metaphor available. That it should go out of fashion was, however, inevitable, since its professors dealt too much in verbal subtleties while neglecting to examine and experimentally interrogate nature through the senses. Had they paid more attention to the sagacious warnings of Roger Bacon and to the example of their great model, Aristotle, modern science might have attained its triumphs by the aid of men who had never lost hold of many priceless philosophical truths. If ever a man was "before his age" it was Roger Bacon, who was not only the warm advocate of experimental physical science, but boldly maintained the true principles of modern Biblical criticism.

Yet his warnings and protestations were of no avail. His contemporaries believed they knew all things knowable, and possessed in one or other of their multitudinous pigeon-holes a solution of every possible problem.

Thus a great opportunity was let slip, and when that change in art and learning, the Renascence, took place, most men turned with a sense of relief to new-born physical science. Such of them as still pursued philosophy, welcomed with avidity a new philosophical departure by a scorner of the old ways—Descartes—who, never having studied the older philosophy, naturally misunderstood it and therefore despised it.

As we before observed, the development of natural science at the Renascence was really an analogous recurrence of antecedent conditions; that is, there was an analogy between its spirit and the spirit of the age of Aristotle. There were, nevertheless, of course, immense differences between the two; but some such return to an antecedent state of things was necessary for subsequent human progress.

Since that day physical science has run a long and prosperous course indeed. By the multitude of problems it has set at rest it has abundantly justified the questioning spirit which is the spirit of all science, and, as we shall see, of philosophical science especially. Nevertheless, physical science, however useful and fascinating, has two notable defects. The first is that it cannot answer the questions men most desire to know. It may aid us in performing our duty, but can never tell us *why*, or even *that*, we should perform it. It can tell us the truth about many things, but can tell us nothing about truth itself. It can afford us good reasons for believing various facts, but not the grounds on which we should believe such reasons. It is essentially superficial, and not fundamental. This is no cause why we should disesteem it, any more than we should disesteem our baker, because he does not know the most recondite principles of biology; or our butcher, because he does not appreciate the bearing of embryology on zoölogical classification. The mode of explaining the universe most popular with physicists is not regarded as satisfactory by many men who feel very deeply the problems of human life. The picture of an infinite aggregate of minute solid balls or microscopic ether whirlpools carrying on a very complex dance, even though intelligence be deemed an accompaniment to some of its more involved figures, is not satisfactory, even as a working hypothesis, to the deeper thinkers among us. But it is the multitude of shallow thinkers who suffer on account of the second defect of physical science. This second characteristic is its tendency, by cheap and easy appeals to the imagination, to stifle the craving of the intellect to know all that can be known of the wonderful universe which on every side surrounds us. We are openly counselled to rest contented with "appearances" (*phenomena*), and to pass our lives as the witnesses of what we know to be but a phantasmagoria, without trying to obtain—what we are told we can never obtain—some knowledge of the painted screen, the pigments, the lanterns and the actors themselves (so to speak) which play off before us the set of dissolving views, the relations between which are all that physical science professes, or even hopes, to be able to give us any information about. There are now many persons, and I am among the number, who think that the time has come for another renascence—a recurrence of conditions analogous to those which existed before physical science became the fashion of the day. It is almost needless to say that the present writer in no way wishes to restore mediævalism, even were such an undesirable miracle possible. All he advocates is a renewed attention to principles and intellectual perceptions which have for a long time been too much lost sight of. He is convinced that these require to be once more keenly and searchingly scrutinized in the light which four centuries of patient and laborious work of physical science have obtained for us.

This first portion of our essay is, in fact, a plea for attention to questions the profound, though generally hidden, effects of which on human life and progress have been unduly—unreasonably—crowded out.

The time, it is believed, has arrived for stimulating the spirit of inquiry in what is for most of our contemporaries a new direction. While carefully gathering up the fruits obtained for us by physical science, we must turn with earnestness and energy to investigate the foundations of all science and of all rational human activity of whatever kind.

Whoever would so investigate must never rest satisfied with any assertion the truth of which is not evident to his reason. He cannot repose upon authority, but must see clearly the truth of each step he takes in reasoning; otherwise the result cannot possibly be a satisfactory one. He must see the truth as to each such step, and know that he sees the truth of it.

This last necessity constitutes the difficulty of the investigation for the great majority of men. We spoke of the need of being willing "to take trouble," and the "trouble" we referred to is the trouble a man often feels when, for the first time, he begins to examine into his own mind and see what his different mental acts are. This trouble is not to be avoided. Every kind of work requires a proper handling of its proper tools. We cannot cut down a tree with a painter's brush, and we cannot understand the value of our different beliefs and convictions save by looking into our own minds and seeing what it is we think and know, and what we are really certain about.

The trouble which at first attends upon this process of introspection is due to a peculiarity of our organization. We are spontaneously impelled to notice surrounding objects, to the apprehension of which the mind applies itself with extreme facility, but we are not so impelled to notice our own various mental states. The child soon learns much about things external to it, but not till long afterwards does it begin to pay attention to its own feelings and thoughts; yet the difficulty we may find in turning the mind inward upon itself can soon be overcome; for the faculty of introspection, like our other faculties, may be strengthened by exercise, and all that is ordinarily needed to perfect it is patient perseverance.

In examining the foundations of all science our inquiry must be, "What is most certain?" As it is impossible for any of us to look directly into any mind but his own, this inquiry must take the form, "Of what things am I most certain?" And this question may be preceded by two others: "Are we certain of anything?" and "Does such a thing as certainty exist?"

Doubt and scepticism are not only legitimate but necessary in science. These are our safeguards against rash assent to propositions inadequately proved. They are doubly necessary for the helpful science of sciences, in studying which we, as before said, should assent to nothing which is not clearly and evidently true to our own minds. Yet here, as elsewhere, there may be exaggeration. It is possible to be so possessed by the spirit of doubt as to forget the existence and legitimacy of certainty.

But, in the first place, we are all certain, as before said, that science advances, and it is obvious that such advance would be impossible if we could not, by observations, experiments, and inferences, become so certain with respect to some facts as to be able to make them the starting-points for fresh observations and inferences as to other facts. Thus, as to the earth's daily revolution, the stratified composition of its crust, and the fact that such strata here and there contain the fossil relics of ancient forms of life—as to such things, educated men are certain. No one will probably deny that we may repose with absolute confidence and entire certainty upon a variety of such assertions.

As to matters of every-day life as distinguished from scientific truths, though we therein generally act on reasonable probabilities, yet certainty meets us at every turn. Thus we are absolutely certain that a door must be closed or not closed; that if having been open it is now shut, some person or thing must have closed it; that we cannot both spend our money and keep it; that we are warm or are enjoying a sweet taste while we are so doing. Such certainty is quite beyond the reach of doubt, and no one denies that while we are actually experiencing some kind of feeling, we absolutely know and may be certain that we do feel it.

But it is necessary at the outset of our inquiry to take very special pains not to make even such assertions hastily. Nevertheless, every man of science must affirm that there really is such a thing as legitimate certainty, and that we are all certain of something; otherwise there would be no certain science of any kind. Blind disbelief is as fatal to science as blind belief, and it is possible for men to get themselves into a diseased condition of general distrust and uncertainty —to acquire a sort of mental falling-sickness. Experience proves that they may bring themselves to doubt or deny the plainest truths, the evidence of their senses, the reality of truth or virtue, and even their own existence. It is necessary, then, distinctly to recognize that there is such a thing as legitimate certainty, not to perceive the force of which is illegitimate doubt. Such doubt would, as just observed, necessarily discredit all physical science. Universal doubt is simply an absurdity—it is scepticism run mad. If a man doubts whether there is such a thing as rational speech, or whether words can be used twice over by any two people in the same sense, then plainly we cannot profitably argue with him. But if, on account of his very absurdity, we cannot refute him, it is no less plain that he cannot defend his scepticism. Were he to attempt so to do, then he would show, by that very attempt, that he really had confidence in reason and in language, however he might verbally deny it. Universal scepticism is most foolish because it refutes itself. If any one were to say "Nothing is certain," he would necessarily contradict himself, for he would thereby, in the very same breath, say that "something is certain," since he says "It is certain nothing is certain." He says, therefore, something which, if true, absolutely contradicts what he affirms. But a man who affirms what the system he proposes to adopt forbids him to affirm, and who declares that he believes what he at the same time declares to be unbelievable, should hardly complain if he is called foolish. No system can be true and no reasoning can be valid which inevitably ends in absurdity. Such scepticism, then, cannot be the mark of an exceptionally intellectual mind, but of an exceptionally foolish one, and every mental attitude which necessarily leads to and results in scepticism of this sort, must be an untenable attitude and a false intellectual position.

Let us see a little further how self-refuting such modes of thought are. Suppose a man were to say, "I cannot be sure of anything, because I cannot be certain that my faculties are not always fallacious," or "I cannot be sure of anything, because for all I know I may be the plaything of some demon who amuses himself by constantly deceiving me." In both these cases such a man would simply contradict himself; for how does he know that "constantly fallacious faculties" or "a demon deceiving in all things" would necessarily deprive him of certainty? Obviously he can only know this because he sees that *it is impossible to arrive at certain conclusions by means of anything uncertain or false.* But if he knows *that* truth—if he is certain of *that*—he must know that his faculties are not always fallacious and that his demon has been unable to deceive him in everything. Universal doubt, then, is manifestly an absurdity.

My object in making these remarks is to enable any of my readers who may need it to get clear of such mere idle, irrational doubts, and to recognize the fact that they already possess an absolute certainty as to *some* things.

We possess mental faculties whereby we can and do arrive at a knowledge of some truth, and if any man professes to doubt or deny their validity, we must be content to pass him by, while calling his attention to the fact that he refutes himself. It is clear, then, that we all do know and are certain about something, as, for example, those persons who read these lines with a view to consider the grounds of certainty may be certain that they are inquiring about the certainty of knowledge.

It being thus clear that we are, all of us, certain about some things, let us see a little more as to what some of them are, and why we must be certain about them. *What are the grounds of our certainty?*

Various theories have been propounded on this subject by a number of estimable persons. Thus it has been said that we can by reasoning attain to a solid support for all our beliefs and convictions. But in order to prove anything by reasoning, we must show that it necessarily follows as a consequence from other truths on the truth of which its own truth depends. Such other truths must, therefore, be deemed more indisputable than the thing they are called on to prove. But it is evident that we cannot prove everything. However long may be our arguments, we must at last come to ultimate statements, which must be taken for granted, as we must take for granted the validity of the reasoning process itself. If we had to prove either the validity of that process or such ultimate statements, then either we must argue in a circle or our process of proof must go on forever without coming to a conclusion. In other words, there would be no such thing as proof at all.

It has, again, been said that indisputable propositions are those which have not been impressed upon us by habit or by any association of ideas, but are what they call "the genuine testimony of consciousness," such as spontaneously arises in the dawning intelligence of the infant mind. It is, to say the least, somewhat difficult to see why such a surprising keenness of mental vision should be attributed to babies; but as such a test of truth is an utterly impracticable one and, like the others here referred to (as we shall shortly see), not really to the point, no more need here be said about it.

Some good persons, finally, lay it down as our duty to select as the truest propositions those which, not having been gained by experience, are called *a priori*, and have been implanted in our intelligence by a benevolent and all-wise Creator. That a judgment is "God-implanted" is a very good reason for accepting it with those who already believe in "an all-wise and benevolent Creator"; but that the test is a practically useless one is plainly shown by the number of books which have been written to refute persons who affirm that we have no sufficient evidence of God's existence or, at least, of his goodness.

Other persons deny that we can discover any indisputable propositions at all, basing their denial on the alleged fact that the whole of our ideas are simply derived from ancestral feelings of not only countless generations of mankind, but of an indefinite number of brute ancestors in addition. But why are ideas and beliefs to be considered less certain and ultimate if they are attained by the help of such ancestral experiences than if they are due to individual impulses?

In fact, one and the same objection must be made to all these different representations. The matters they refer to are very interesting, but the problem we have to solve is one entirely independent of them and has nothing to do with the problem concerning the *origin* of our judgments. Valuable as such inquiries are for the study of the human mind, they are quite out of place in an inquiry as to what judgments are evidently and supremely certain. This last inquiry refers to the grounds of belief which any judgment may exhibit in and by itself—to a criterion of its truth—and not at all to the causes which have produced it. Yet there are philosophers who have been so busy in trying to find out how different propositions have come to be believed that they have neglected the more fundamental inquiry as to *why* they should be believed — what

grounds of certainty they exhibit. By the
"grounds of certainty" which any judgment
can show, it is not, of course, meant anything external to it. Such a meaning would
imply a proof of the judgment, and would
involve us in an endless and resultless series
of arguments, as already pointed out with respect to reasoning. The only ground of certainty which an ultimate and supremely
certain judgment can possess is its own
internal self-evidence—its own manifest certainty in and by itself. Such certainty we
possess when we recognize that we feel pain
or that we have a sensation of sound or a
taste of sweetness. It is manifestly as superfluous as impossible to try and prove such
things; their evidence is direct and immediate, and whatever may be the origin and
causes of such perceptions, they manifest
their existence by their own internal evidence. The slightest reflection will show
that it is impossible that we can be deceived
about having a feeling while it is being absolutely felt. Similarly all proof, as all reasoning, must ultimately rest upon truths

which carry with them their own evidence, and do not therefore need proof.

But some readers may be startled at the suggestion of believing anything on "its own evidence," and may fancy it is equivalent to a suggestion that they should believe something *blindly*. This startled feeling is, I believe, due to the circumstance that the greater part of our knowledge has been gained, not by our own observation, but by testimony or else by inference.

We ordinarily ask for some "proof" with regard to any remarkable statement, especially if it be one which carries with it important consequences. No truths are brought home to our minds more forcibly than are the truths of Euclid, and these have been rigidly demonstrated to us by the clearest reasoning. From such habitual connection between (1) reasoning and (2) our apprehension of new truths, we have acquired a feeling that to believe anything which cannot be proved is to believe *blindly*, and have therefore acquired a tendency to distrust whatever may be above or beyond proof.

We are too apt to forget, what, on reflection, is obvious enough, namely, that if it is not blind credulity to believe what is evident to us by means of something else, it must be still less blind to believe that which is directly evident in and by itself.

The most certain of all judgments, then, must be those which require no proof and are self-evident. But is it within the bounds of possibility that the most certain and ultimate truths could have any better criterion or ground of belief than they in fact thus have?

All criteria or tests of the truth of judgments must be one of two kinds:

Every such test or criterion must either (1) reside in the judgment or perception itself, and so make it luminously self-evident, or it must (2) reside in something external to it. Now any external criterion, however striking and perfect it might be, could only be appreciated by us through our perception of it and our judgment about it. If a proposition appeared suddenly written up on the clouds or on the face of the full moon, we could not recognize such a proposition as true until after having examined and carefully considered much evidence about it. Our first impression would of course be that we were the victims of an hallucination, and then the question of the possibility of common hallucinations would have to be considered. Ultimately and at last, if the truth of such a proposition were accepted, it could only be accepted because we perceived that our ultimate judgments about it were self-evidently true. In fact, it would, for us, at least, exclusively *repose upon* that very *self-evidence*, objections to which have been here suggested. By no external criteria, then, neither by such as the absurd one just imagined nor by any other, could we be made any better off than we now are. We could but have self-evidence, after all, as our ultimate criterion. It will be plain, on reflection, that nothing external—no common consent of mankind, common-sense, or testimony—could ever take the place of an ultimate criterion of knowledge, since some judgment of our own mind must always decide for us with respect to the existence and value of such criteria. Self-evidence, then, is the necessary and the only criterion of truth. The principle of evidence is one which is really ultimate, and must be accepted under pain of either futile reasoning or complete intellectual paralysis. It is, however, necessarily incapable of demonstration, since it depends on nothing else. We all of us assume it as a criterion unconsciously, and it is confidently acted on by every one who reasons. But when we ponder over the matter it becomes clear that what we have thus done, through the spontaneous activity of our intellect, has been done *most* reasonably. Did we not adopt it, we should not only be utterly unable to think logically, but should be plunged into the most utter and most absurd scepticism. On the other hand, by recognizing that criterion for what it must be and is, we gain a secure foundation for our knowledge and are enabled to make progress in science. Our mental condition is transformed from a hopeless chaos to an orderly cosmos.

We end this preliminary part of the present essay, then, by claiming to have shown the attentive reader that his own mind already possesses absolute certainty about some things, and that the declaration of his intellect is that things which are clearly seen to be evident in and by themselves possess the greatest certainty which it is possible for the human mind to have. If any one is so unfortunate as not to be able to see this clearly and not to see that there is such a thing as certainty with respect to any matter, then he had better read no further, and content himself with simple matters, the toils and the pleasures of everyday life, without a thought beyond. Fortunately it is not necessary to be a wise man in order to be a good and useful one in plain and simple ways of goodness and utility.

But for healthy, normally constituted minds that can appreciate the necessary conditions which appertain, for us, to all fundamental truth, it must be evident that we need three things in order to enter upon the investigation of the ultimate foundations of all physical science and all rules of conduct. These three things are: (1) perceptions of certain general principles, (2) perceptions of certain facts, and (3) perceptions of the validity of certain arguments. Without a knowledge of certain general truths we could not argue, without a knowledge of certain facts all our reasoning would merely concern ideas and would have no relation to reality, and without some criterion of valid reasoning we could never arrive at any conclusion, and all argument would be vain. Since, however, men of all schools argue, it is plain they do not think all argument vain, and it is no less plain that they must accept some facts as true and rely with confidence upon those principles to which, in their arguments, they make appeal. In the next part of this essay we shall directly apply ourselves to consider some of the truths which are implied and confidently relied on by all followers of physical science.

Part II

THE object of the first part of this essay was to bring home to its readers that certain very practical questions have been unreasonably crowded out in that unconscious competition for the attention of mankind which may be called "the struggle for life" of the arts and sciences.

Considerations also were therein proffered which seemed to prove that some correct study of "the helpful science" is incumbent on all who desire to lead a thoroughly well-ordered and entirely reasonable life.

It was also pointed out that we all possess absolute certainty about some things, and that the result of self-questioning is to show that nothing can be more certain than whatever is self-evident in and by itself. Therefore, all sciences, together with all ethical precepts, ultimately repose on certain truths seen to be evidently true—an evidence than which there can be no better.

The truth of whatever is true and the evidence of whatever is evident can, of course, only be made known to us as being clearly true and evident by means of "thought"; as, for example, when the reader of this page turns his mind in upon himself and recognizes what he is doing and that he knows it. Not to be certain about such an evident truth is to be in a mentally diseased and more or less intellectually paralyzed condition.

The author craves indulgence for such seemingly trivial assertions, but it is not his fault that he has to make them. The ingenuity of successive metaphysicians, each trying to be original and obtain renown by some new absurdity, has made it necessary for any writer who aspires to be helpful to make very sure of the ground he proposes to build on, and that every stone of the building he begins to erect should be firmly and solidly set.

But here I may be met with the objection,

"Your ground is not sure; you represent all truth as being known by means of 'thought'; but the railway I travelled by yesterday was not 'a thought,' and the dinner I hope shortly to partake of will, I trust, be made of more substantial material. Thank God, I have my senses about me, and I shrewdly suspect that, I begin to feel sensations a considerable time before I begin to think about them. I abide by my experience; I know my feelings and sensations, and I am much more inclined to regard all your fine 'thoughts' as really built up of them, than to think the cramp in my leg which woke me last night was 'a thought' or an amalgam of 'thoughts.' I therefore demur to your system altogether, and at the very outset."

There are not a few men who, like my supposed opponent, would assign to our sensitive faculty the dignity of acting as our supreme and ultimate test of reality and truth. But the following considerations will, it is believed, suffice to show that in this such men are greatly mistaken. It is the intellect which is alone supreme, and this not only in judging about abstract or mental matters, but also in judging of matters of which the senses take cognizance—whether it be railway trains, the courses of a well-ordered dinner, or anything else.

I am not one of those who doubt about the real existence of an external world, or about the certain information our senses can give us concerning that world. Nor do I deny—for I confidently affirm—that sensation precedes thought, and that only after a long course of sensations felt does the infant begin to think. But, as pointed out in the first part of this essay, the question is not as to the mode of origin of the intellect's activity, but as to the value of its declarations when mature. He would not be a very wise man who in business matters should doubt the accuracy of his calculations because he could not have made them when he was a week old! But even with respect to the material objects about us which can be submitted to the test of our sensitive faculty, the last word, in all cases of doubt, rests with the intellect and not with the senses. Let us suppose that we are performing some chemical experiments to find out the properties of some newly discovered compound. Since in such a case we appeal directly to our senses for information, it might seem that our ultimate criterion must be our sensations, and that their declarations must always be supreme. Yet in fact such is not the case. The enormous value of our sensations is manifest, as well as that the information we receive through them is indispensable. Nevertheless, it is the judgment of the perceiving mind which alone gives them value, although its action ordinarily takes place unnoticed. Let, however, some distrust arise as to the results arrived at, leading, perhaps, to a careful repetition of experiments; then, in the last resource, when we have done observing and experimenting, how do we know we have obtained the results we may have obtained, save by the intellect? How are we to judge between what may seem to be the conflicting indications of different sense-impressions? Nothing could be more foolish than to undervalue the testimony of the senses, and the senses are truly *a* test and cause of certainty, but they are not *the* test of it. Certainty is not in sensation, though sensation is so constantly our means of acquiring it. Certainty belongs to thought, and to thought only. Self-conscious, reflective thought, then, is our ultimate and absolute criterion. It is by thought only — by the self-conscious intellect — that we know we have "feelings" at all. Without that we might indeed feel (since we often come to know we have had feelings of which we were at the time unconscious), but we could not *know* that we felt. Our ultimate court of appeal and supreme criterion is the intellect, and not sense; our last appeal is and must ever be to a perception or "*intuition*" of the intellect.

But not only in such appeals to our own mental experience is there a supremacy of "thought" over "sense," but every one of our perceptions—be it of a dog or a dollar—contains what is altogether beyond sense. To perceive a dog implies a perception of "unity," and therefore of "number," of "existence," of "distinction" (as distinct from creatures *not* dogs), of "reality" (as not being a mere imagination), and of "truth."

The profoundly irrational system which would feign persuade us that ideas are nothing but sense-impressions—received, associated, remembered, etc.—and that our ultimate appeal is to the senses, may be justly spoken of as "*Sensism*" and its advocates as *Sensists*.

The opposite system, here vindicated, may (since it recognizes the intellect as supreme over sense) be rightly termed *Intellectualism*, and its vindicators—as alike opposed to all forms of materialism or idealism—named *Intellectualists*. It is quite true that multitudes of objects of all kinds existed ages before our intellectual activity began, but such things were practically non-existent for us until our intellect began to think about them, and so furnish us with a basis for an incipient degree of knowledge. We can feel, and originally did feel, without exercising our intellect. But until such sensuous experience comes within the grasp of the intellect—comes to be an object of our direct thought—it cannot constitute "knowledge."

Space cannot be here afforded for any lengthened demonstration of the essential difference which exists between sensation and thought. It must suffice to show, by one or two more examples, how, by the help of sensations, the mind can conceive of what is altogether beyond sensation. Thus, for example, we often refer to some past " experience," and the idea is a sufficiently familiar one. Yet that idea cannot possibly be a faint reproduction of past feelings, for " experience " is something which was never felt at all. By receiving, or obtaining, over and over again, feelings of the same or of different kinds, we may find it easier to feel them; or they may become more pleasurable to us or (as is too often the case) less so. But to undergo such changes of feeling, and to obtain the idea " experience," are two very different things. Again, we can all form an idea of the action of our eyes in seeing (our act of sight), yet that act of seeing was never itself felt, nor can the idea be decomposed into mere feelings—there is infinitely more contained within it. We may have a certain feeling in our eyeballs while looking, but even if we could feel (which we cannot) every minute action of every part of the eye's and the brain's complex mechanism, such feelings would be no "idea of the act of seeing." Among the constant experiences of our daily life are our perception of different shades of color, and different feelings have gone along with them. Of "color," however, we have never once had a feeling, yet we have a clear idea of it, and often speak thereof.

We have certainly another idea which was never felt, and that is our idea of "nothing," or "nonentity." It is very certain that past sensations can never account for *that* conception, which is, nevertheless, one commonly enough employed. How often do we not hear such expressions as "It is worth nothing," or "There is nothing in it!"

So much for the fact that ideas have an extension and validity beyond the domain of our mere sensitive faculty. Nevertheless, I am far, indeed, from regarding ideas as everything—which is "Idealism."

It is utterly impossible in the course of this essay to enter upon any examination of that system of thought which originated with Bishop Berkeley, although I shall have some words to say about it later on. A disbelief in—even a doubt about—the real, independent existence of the material world is, in my eyes, a form of insanity. That, for example, the numerical relations of things—the facts concerning their numbers (that, *e. g.*, oranges on a plate are three and not five)—are matters absolutely real and true *in themselves*, and are not dependent on my faculties or the thoughts and perceptions of any other man, is, for me, unquestionable. That the bodies about us are really extended—that the quality extension is a real one—is also, for me, unquestionable. When Dr. Johnson kicked a stone by way of refuting idealism the action seemed absurd, and has been long a subject of ridicule. For all that, I strongly suspect that the existence of a direct perception—an intuition—of the real extension of seemingly extended things was the truth he intended to assert by that rough-and-ready procedure.

But it is time to revert to the consideration of some of those truths which are implied, and confidently relied on, by all followers of physical science—a consideration which it was said, at the conclusion of the first part of this essay, should form the subject of its second part.

Those truths, as before pointed out, relate to three orders of truth: (1) general principles, (2) particular facts, and (3) the process of reasoning.

Now every reasonable man, and surely, above all, every practical man, must need some especially solid foundation as a starting-point, and, therefore, I will select from the second of the three foregoing categories one special and particular fact as an absolute starting-point in the investigation and justification of "the helpful science."

Let us suppose that certain definite observations and experiments have been carried on—such as those which have been performed by M. Dareste in order to find out what physical deformities in the embryo of the chick may be produced by certain exceptional modes of treatment of the egg. Now there is one supremely important truth which is implied in our certainty of the result of any such experiment, whatever that result may be. Unless we can be sure that it was we who both began the experiment and also witnessed its conclusion — that there had been no change in our personality while experimenting—such conclusion could not be confidently relied on by us. In other words, the truth implied in every scientific experiment is the certainty of our knowledge of our own continuous existence during its performance.

But here some of my readers may very naturally exclaim, "I am not going to waste my time in reading assurances that I know I exist. I am not an idiot!" And, indeed, when some one is very sure about anything he will often say, "I am as certain about it as I am of my own existence." It is no wonder, then, that some persons should feel disinclined to consider reasonings about a matter they have never doubted in the least, even though they may know that others have professed to doubt about it. "To doubt one's knowledge of one's own existence," they may say, "is even more absurd than to believe that one's limbs are made of glass." Nevertheless, we once more beg our readers to exercise yet a little patience still, for the question is one of extreme importance. It is denied by all the modern followers of Hume—such as Herbert Spencer, Huxley, Mill, Bain, etc.—that we have, or can have, supreme certainty about our own existence. Such certainty, they say, we can only have with respect to our actual present feelings, or what they call "states of consciousness." They do not, of course, deny the certainty of the momentary feeling of existence which any one may have; what they deny is that we can have any certainty that our existence continues on, substantially the same, during all the various changes of feeling we successively experience.

Such a "continuous self," it is affirmed,

we can never know apart from our various "feelings" or "states of consciousness." We can never know our own self, they tell us, unmodified; that is, as existing what they call *absolutely*, or, in other words, by itself. Therefore, they argue, we can never know our own existence with the greatest certainty possible.

But, in fact, there is nothing we can perceive which exists apart from everything else or "absolutely"—as it is, in my opinion, very absurdly called. Everything which exists always exists in some state or condition, and stands in some definite relation to other things. Small wonder, then, if we do not know things in a way in which they never do and probably never can exist; that is, "unmodified" and "unrelated." We can know nothing by itself, for the very good reason that nothing exists "by itself." Therefore, it is not at all wonderful if we only know "ourselves" in connection with our various "feelings," or vice versa.

It is quite true that we never knew our own existing being, alone and unmodified;

but then we have never for an instant so existed. Our knowledge of ourselves is, in this respect, just like our knowledge of anybody else. Probably some of our readers have known the late highly distinguished botanist Professor Asa Gray; yet, if so, they never knew him except in some " state "—either talking or silent, at home or away from home, examining a plant or not examining one, with his head covered or with it uncovered—and this for the very good and obvious reason that he never did, or could, exist for one moment save in some "state." But this need not have prevented Professor Asa Gray from being very well known, and the same consideration applies to our knowledge of ourselves. If the reader will reflect, and consider what is his primary, direct consciousness at any moment, he will find it to be neither a consciousness of a "feeling" nor one of "his continuous existence," but a consciousness and perception of doing something (as, for example, reading this essay), or else of having something done to him, by some person or thing—it may be being supported by a chair on which he is sitting. But this consciousness he will find to be always accompanied by some "feeling" or other, and also by some "sense of his self-existence." He (it may be he who reads this) can, indeed, whenever he likes, make himself explicitly aware either of his "feeling" or his "sense of self-existence."

He can do this by turning back his mind upon itself for that purpose, and then he will be able to say to himself either, "I have the feelings which attend holding and reading a book on the helpful science," or he may say to himself, "It is *I* who have these feelings." But this is not a natural primary act, but an act of reflection; that is, a secondary act. No one, when he begins to think, adverts either to his "present feelings" or to his "continuous existence." No one begins by perceiving his act of perception a bit more than he begins by expressly adverting to the fact that it is he himself who perceives it.

It is only by reflecting on the direct spontaneous perception of the mind that we can see (by such a reflex or secondary act) that our perceptions and feelings *are* perceptions and feelings, or that it is truly *we* who perceive and feel. Let us suppose two men to be engaged in a fencing-match. Each man, while he is parrying, lunging, etc., has his "feelings" or "states of consciousness" and knows very well that it is he who is carrying on the struggle. Yet it is neither his "state of consciousness" nor "the persistence of his being" which he directly regards. What he directly regards is what he is doing and what is being done to him in attack and defence. He can, of course, if he likes, direct his attention either to the feelings he is experiencing or to his underlying continuous personality. Should he do so, however, a hit from his adversary's foil will be the probable result.

But to become aware that one has any definite "feeling" is an act of reflection *at least* as secondary and posterior as it is to become aware of the "self" which has the "feeling." We say "at least," but I believe that of the two perceptions — (1) that of "feeling," and (2) that of "self"—it is the "self" which is the more prominently given in our primary, direct cognitions.

I believe, nay, I am sure, in my own case, that a more laborious act of mental digging is needed to bring explicitly before the mind the "feeling" implicitly contained in any perception than to bring explicitly before the mind the sense of "self-existence" implicitly contained in any such supposed perception.

Men continually and promptly advert to the fact that actions and sufferings are their own (they will quickly claim the merit of the former, or cry out against the supposed undeservedness of the latter), but do not by any means so continually and promptly advert to the fact that the feelings they experience are "existing feelings."

Therefore I am profoundly convinced that one of the greatest and most fundamental errors of eccentric writers and lecturers of our own day is the mistake of supposing that we can know our "mental states" or "feelings" more certainly and directly than we can know the "continuously existing self" which has those feelings.

To make my contention still plainer, let us suppose that a man of ordinary susceptibility has received a slap in the face. What is his immediate, explicit perception? Not that a certain "state of consciousness" exists, nor that there is an "enduring self" which has become newly modified. His direct perception is that he has been struck, and different "feelings" will accompany that perception according to the circumstances of the case. He may then, if he pleases, either explicitly examine his "feelings," or explicitly consider "himself" as affected by what has occurred; and in each case he will by so doing hold up, as it were, to his mind's eye the "feelings" and the "self," and regard them by that second application of the intellect which I have before termed "reflex." But he cannot so examine his "feelings" without a perception that they are his own; nor examine "himself" without a perception of the more or less vivid "feelings" which have just been aroused in him.

I am at this very moment writing. I feel the pen and I feel the motions of my hand and arm. But ordinarily, when writing, I no more advert to such "feelings" than I advert to the feelings in and the movements of my legs as I am running up-stairs. It is plain that we do not so advert; for as surely as our attention is so directed, our writing movement is hampered in the one instance, and a stumble on the staircase is very likely to occur in the second. Much less inconvenience ensues from turning the mind inward (while writing or running up-stairs), and recognizing the fact that it is we ourselves who are in the act of doing either one or the other of these things. Thus here, again, we may recognize the fact that of the two certainties, the certainty of our own existence is more easily attained than is the certainty as to what the nature of the various feelings which accompany the actions may be, whether we are writing, running up-stairs, or whatever we may be doing.

This truth is, I believe, the most important and fundamental of all the truths our minds can give us any information about—not on its own account, but on account of the consequences which follow its distinct recognition. We have but to turn our minds inward and advert to what our consciousness tells us, in order to be able clearly to see that the fact of our own existence is a truth which carries with it its own evidence, and is absolutely certain in and by itself.

It is not, however, a momentary existence which it suffices for us to recognize. If our experiments have (as supposed) been directed to ascertain any kind of scientific truth, it is evident that we must be sure that our existence has persisted from the beginning of such experiments till their conclusion; otherwise all scientific certainty, as before observed, would be at an end. We may be sure, then, that we can know not only our actions, sensations, imaginations, reminiscences, emotions, perceptions, and conceptions, but also our own substantial and continuous personal existence. The fact that we possess this knowledge, however, implies another most important truth, namely, the validity of our *faculty of memory*.

And, indeed, this is necessarily implied in the pursuit of physical science. We have spoken of M. Dareste's experiments with eggs, but it must be plain to every one that no result could be arrived at, let such experiments be ever so carefully performed, if we could not trust our faculty of memory.

It is true, of course, that we may make mistakes as to particular things, and defects of memory, which occasionally occur, are very various—affecting only certain parts of speech or other limited department of the mind. But such exceptional phenomena do not tell against the fact of the general trustworthiness of memory.

It is also plain and obvious that the trustworthiness of memory is implied in our knowledge of our own existence, since we could never know either what (1) our recently experienced "feelings" or (2) our "perceptions of self" have been, save by the aid of memory. Therefore, the certainty we have as to the one or the other of these, carries with it a certainty as to our act of memory.

But what is memory? Evidently we cannot be said to remember anything unless we know that the thing we so remember has been present to our mind on some previous occasion. A mental image might present itself to our imagination a hundred times, but if each time that it so presented itself it seemed to be something altogether new and unconnected with the past, we could not be said to *remember* it. It would rather be an example of extreme forgetfulness than of memory.

By asserting the general trustworthiness of our faculty of memory we do not, as before said, mean to deny that mistakes are often made. Nevertheless we are all of us certain as to some past events. Probably every reader of this essay is absolutely certain that he was doing something else before he began to read it. Memory gives us as much certainty concerning *some* portions of the past as we can have with respect to some portions of the present.

If we could not trust our faculty of memory all science would be, for us, a mere present dream. But the veracity of this faculty is a self-evident truth. It can never be proved. There can be no such thing as a *proof* of it, because we cannot argue at all unless we already trust it. Yet, marvellous to say, Professor Huxley has declared that we may trust our memory because we learn its trustworthiness *by experience!* This declaration amounts to a demonstration of the Professor's incapacity in matters of philosophy. Surely never was fallacy more glaring! How could we ever gain experience at all, unless we trusted our memory in gaining it? Professor Huxley has thus said, in effect: "You may trust your present memory because experience has confirmed it, while you can only know that it has confirmed it by trusting your present memory!"

Memory, as we shall see, directly performs a yet more wonderful office than any I have yet referred to. Before, however, stating what it is, I think it well to explain the only two technical terms of which I propose to make use.

These two terms may be represented, to a certain extent, by the familiar words, "*facts*" and "*feelings*." The technical terms which correspond therewith respectively are: (1) things which are "*objective*," and (2) things which are "*subjective.*"

Every "feeling," "thought," "sensation," or other "state of consciousness" present to the mind of whoever is the subject of it, is spoken of as being "subjective." It is a thing which pertains to the subject—to the mind which feels or thinks. The whole of such experiences taken together constitute the *subjective world* or the sphere of *subjectivity*.

On the contrary, everything whatever which exists externally to our present consciousness or feelings, is spoken of as being "objective," and all that is thus external to the mind constitutes the *objective world* and is the region of *objectivity*.

It is the world of real *objects* which occasions thought or feeling, as opposed to the subjective modifications of him who thinks or feels.

Everything which is *subjective* pertains to the *self* or ego during the time in which that "self" is feeling or thinking.

Everything which is objective is external to the self which is feeling or thinking, so that all states, even of the "self" or "ego," which are anterior to the time when that self or ego feels, is also objective — an *object* of thought, indeed, but not the thought or feeling of the thinking subject—not *subjective*.

Now the wonderful office above spoken of as being performed by memory concerns this distinction between objectivity and subjectivity, and is, in fact, the bridge which exists, in our being, between the two.

It is memory which enables us to get, intellectually, outside our present selves and our present feelings and sensations, in a way the truth of which no sane man can question.

For memory informs us with absolute certainty as to some events of our own past history. But such events are beyond our present experience; therefore they possess a truth which extends back beyond any present feeling. They are realities which are anterior to our existing feelings, and they are, therefore, *objective*.

Thus memory, inasmuch as it reveals to us part of our own past, reveals to us what is " objective," and so actually introduces us into the realm of *objectivity*, shows us more or less of " objective" truth, and carries us, as before said, into a real world, beyond the range of our own present feelings.

The power which memory possesses of thus lifting us, as it were, out of our present selves, and showing us facts which otherwise we could never know, is certainly a most wonderful power; and the plain fact that if we admit the validity of memory in the least —if we can be certain of even one single thing which took place only a few hours ago —it is indisputable that we can, and that we *do*, learn *real objective* truth, and can be certain of more than mere " appearances," or " our present feelings only." The fact that we can certainly know these two truths: (1) that we can really know our own existence, and (2) be certain of some facts anterior to our present feelings, is a truth fruitful in far-reaching consequences. The reader may, perhaps, at first be disposed to consider the two truths here pointed out as being too obvious and trivial. But it would be a great mistake so to consider them, as will be plainly seen in the concluding portion of this essay. They carry with them, indeed, tremendous implications.

Such being the case, it will be well for the student of "the helpful science" here to pause a little and well consider these two truths, so that he may see they are absolutely certain ones, though necessarily incapable of proof, while they carry with them their own evidence. He may also reflect that the real validity of all science hangs upon them, although such dependence is rarely adverted to. The veracity of our faculty of memory, and our knowledge of our own existence, are both implied in every scientific observation which we regard as certain, and in every mercantile transaction in which we have entire confidence.

We can know the result of such observations and experiments (as before said) and we can carry on such mercantile transactions only through memory, by the help of which we are, each of us, enabled to unite the past with the present and say "I am." These two words have an immense significance for any one who will carefully ponder over them. They signify that he who utters them intelligently recognizes certain past acts as his own acts, and that a continuous unity (himself) has endured, essentially the same, for a longer or shorter time and has had more or less varied experiences. He who utters them also thereby indicates that he has the power of knowing at least one objective existence which his senses cannot perceive.

Such must be the case, because our senses can only feel what is present to them; they can never feel the past. The very fact of our feeling anything shows, with absolute certainty, that something is actually present which occasions that feeling. But it is clear to every one that his intellect can, by the help of memory, know with certainty something which is far from being present here and now—namely, some event of his past life. Similarly he is enabled to perceive his own continuous existence—which is certainly a thing which cannot be felt.

Our body can, of course, be felt as often as we like, in different ways at once and as long as we choose. Nevertheless, each time we feel it we can but experience the present feeling, and without memory and without reflex acts of the intellect, we cannot know that our body has, and has had, a continuous and enduring existence. It can never be "felt" *as enduring*, although by the aid of repeated sensations it can be recognized by our *intelligence* as enduring. But the intellect, aided by memory, can know very well, by itself, and directly, that it has an enduring permanence, and that the thought of the day before yesterday was its own thought. It can know this with a degree of certainty which it is impossible to attain to with respect to anything else. To doubt the continuous existence of our body would be absurd indeed and a sign of lunacy; but to doubt the continuous existence of the intellect, while illuminated by memory as to some of its past acts known with certainty, would be infinitely more absurd still!

This power of memory, however, is so wonderful, and the consequences it carries with it are so profound, that it is hardly surprising that some persons, in the interests of a philosophical system essentially irrational (as we shall see later on), should have attempted to impugn its validity. But, as we have seen, it is simply impossible for them to do so without contradicting themselves and committing intellectual suicide by falling into a fatuous system of universal scepticism. The self-evident truth that our faculty of memory is a valid faculty, is absolutely necessary to the full recognition of what is, for us, the first and most certain of all facts, namely, the fact of our own existence.

If I have insisted on these truths with some tediousness of iteration, it is because the whole of my future contention depends upon them. My readers will, I trust, therefore pardon me, as unless the certainty of what is contended for in this second part of my essay were made plain and indisputable, "the helpful science" could never be established.

The certainty of these two preliminary "facts" being now clearly seen, we may next proceed to consider one of the two other things I before represented as constituting the foundation of all science and art: (A) "perceptions of certain general truths," and (B) "perceptions of the validity of certain arguments."

It may, perhaps, be better to begin with the last-mentioned category and consider it in relation to what we said in the first part of this essay: the assertion that "the value of science and the fact of its progress are unquestionable."

Physical science, as we all know, advances by means of careful observations and by experiments. But, as we said also in the first part of this essay, not only observations and experiments are necessary, but inferences are necessary also. It cannot get on without them. Therefore the mere fact of scientific progress suffices to show that we must place confidence in the process of reasoning—in that peculiar perception of the mind which we express by the word "therefore." When we use that word we mean to express by it that there is a truth, the certainty of which is shown by the help of other things which have been previously ascertained to be true, and whose truth involves, and carries with it, the truth of the thing they are called in to prove. This is the process of deducing one truth from other truths previously known. It is the process of *deduction*, and every one who reasons shows by so doing that he thinks it possible to draw valid inferences, and implies that he himself has confidence in the principle of "deduction," and recognizes the force of that unique word "therefore." He implies that there is at least one truth the certainty of which is shown through the help of facts or principles which themselves are known to be true.

The process of deductive reasoning has long been familiarly illustrated as follows:

All men are mortal (major premise).

Socrates is a man (minor premise).

Therefore Socrates is mortal (conclusion).

This form of reasoning (consisting of two premises and a conclusion) is, as doubtless all my readers well know, termed a "syllogism."

Now it is sometimes objected to deductive reasoning—to the syllogism—that it really teaches us nothing new, all that is contained in the conclusion being already contained in the premises. Opponents of the syllogism say: "Whoever has said 'all men are mortal,' has thereby also already said that 'Socrates is mortal.' The so-called 'conclusion' is therefore but a repetition of part of the major premise, 'all men are mortal.' Here then we really have no inference at all, but merely a restatement. We do not in truth 'conclude' that Socrates is mortal, but we only say over again, with the mention of his name, what was said before without the mention of his name."

I have wished to draw out this objection to the syllogism as distinctly as possible, because most important consequences follow from the acceptance or rejection of the validity of the deductive process.

There is really no force whatever in the above objection. Such force as it may appear to have simply arises from ignorance on the part of the objector of the great difference which exists between *implicit* and *explicit* knowledge. To cause a knowledge which we only possess "implicitly" to become "explicitly" present to our minds, may often be to practically give us a knowledge of something of which before we had no available or conscious knowledge at all.

Let us suppose that a youth has learned by heart the characters which respectively distinguish the four classes of back-boned animals—beasts, birds, reptiles, and fishes—but that he has seen and knows very little about specimens of different kinds. It would not be wonderful that such a youth should take a porpoise to be a fish. But his teacher might remind him that all creatures possessing certain characters of brain and heart were beasts; and he could thereby come to see that the creature he took to be a fish must, since it has those characters, be a beast. Referring again to the characters of the class of beasts, he might further exclaim, "This fish-like thing when alive must, as being really a beast, have had warm blood." Of course his conclusion would be quite right, and so such inferences would have supplied him with knowledge which he certainly did not possess before.

So great indeed is the difference between explicit and implicit knowledge that the latter may not deserve to be called real knowledge at all. No one will be so absurd as to affirm that a student who has merely learned the axioms and definitions of Euclid has thereby obtained such a real knowledge of all the geometrical truths the work contains, that he will fully understand all its propositions and theorems without having to study them. Yet all the propositions, etc., of Euclid are *implicitly* contained in the definitions and axioms. Nevertheless, the student will have to go through many processes of inference by which these implicit truths may be explicitly recognized by him before he can be said to have any real knowledge of them.

Of course, in the very rare instances in which the major premises express a truth which has been arrived at by an examination of every fact referred to in it—a process which is known as a "complete induction"—there is nothing implicit.

Thus if we knew, with absolute certainty, that *every* man, woman, and child in some Indian village was a leper, then to say that a man came from that village would be equivalent to saying *explicitly* that he was a leper; there would then be no evolution of an implicit into an explicit truth. Such cases are, however, most rare. No one can pretend to know by examination that all the radii of a circle are equal by a complete induction—by an examination of every existing circle. Similarly, if we are shown a triangular figure and are asked, "Are its angles equal to two right angles?" we shall not be able at once to answer the question by directly inspecting the figure. If, however, we already knew that the angles of every triangle are together equal to two right angles, then we should be able at once to infer the truth, and to say that in so far as the figure approximated to an ideally perfect triangle would its three angles approximate the two absolutely perfect right angles. We should arrive at this truth mediately, and reach the conclusion by the help of major and minor premises. A very great part of the knowledge we acquire throughout our whole lives is acquired in this indirect way by the help of that mental process which is expressed by the word "therefore."

But we have no special reason to be proud of that word, since it implies that we are compelled to get at truth by a very roundabout process. Were our intellect of a much higher order, it is conceivable that we might be able to see equally well, and at the same time, all those truths which a proposition may contain *implicitly* as well as *explicitly*. In that case, of course, we should be saved the trouble of any process of inference. The truths we now have to gather indirectly would then be directly evident to us, just as our own actual mental activity is evident to us. Having, however, the imperfect nature we have, we must be content with the more laborious, though practically sufficient, process of inference or reasoning. We must be content to gain actual knowledge from implicit truth by placing propositions side by side, and so evolving explicit truth by such process properly performed. To see to the proper performance of this process is the task of logic, which is at once a science and an art. It is a science, in so far as it reveals to us the laws which regulate human thought; it is an art, in so far as it teaches us how to proceed in order that our inferences may be valid and true.

Reasoning, then, is an indirect process of attaining truths, and one which, when properly carried out, we may entirely trust. Nevertheless, we can already see that it is not the highest kind of act of which our intellect is capable. Such highest act is that by which we perceive—or *intue*—truth directly and without adventitious aid. We perceive such a truth in the direct perception of our own activity, and in our apprehension of those principles and necessary truths which will occupy us in the next portion of this essay.

The three subjects which have here occupied us—(1) our perception of our own existence, (2) the validity of the faculty of memory, and (3) that of the process of inference—are alike truths incapable of proof, and ones which carry with them their own evidence. If any one were to deny that the process of inference is a valid process, he would be landed in absolute scepticism, and would therefore commit intellectual suicide. For if that process be worthless, then all argument must be useless and vain. More than this: not only must all reasoning addressed to others be vain, but the silent reasoning of each man's own mind must be vain also. But this amounts to an utter paralysis of the intellect. It is, indeed, as we said before, scepticism run mad.

In conclusion, I would invite my readers

to note that, in recognizing each of these three truths (our mental continuity, the trustworthiness of memory, and the validity of reasoning) we have passed beyond what is to be perceived by our senses—*beyond* things physical. We have thus, without, I hope, any undue strain upon my readers' minds, distinctly entered the domain of ultra physical truth, or *meta*physical knowledge. Such knowledge, unnoticed though it be, lies at the root of all that is either "good" or "true" in human life.

Part III

IN commencing this third part of my essay on the Helpful Science, it may be well to address a few words to any of my readers who may feel some disappointment at not having as yet found some or other help they may have hoped to find in it. Of such I would ask a little further patience, and beg them to reflect that no structure can be safely reared save on a foundation firm enough to uphold it. We have in the last part laid three solid foundation-stones, and we must lay yet two more before we can begin any solid work of construction.

I would also beg such readers to note that though at first starting they may not have felt the slightest doubt about their own existence, the general trustworthiness of memory, or the possibility of arguing logically, they may now possess not only a full and *explicit* certainty about these truths, but also a knowledge, by reflexion, that they can and do possess it; as, also, that these are fundamental verities carrying their own evidence with them—an evidence than which none can even be conceived of as more satisfactory and certain. There is yet another truth which we shall hereafter see to carry important consequences with it, and that is our recognition that we are beings possessing intellects capable of pondering about such verities and seeing the absolute certainty of them. But, indeed, we are already in a position to appreciate, to some extent, the real helpfulness of the helpful science of metaphysics; for we have seen that it is the *implicit* knowledge of the three metaphysical certainties (to the establishment of which the second part of this essay was devoted) which has alone rendered the progress of physical science a possibility. The utility of an agent does not necessarily depend on the recognition of that agent's utility, and still less on the causes which have made it useful. Atmospheric and marine currents carried on their various agencies, so helpful to human life, for ages before their utility was recognized, and, *a fortiori*, before the causes which produced them were satisfactorily elucidated. Therefore, the title of "the Helpful Science" would none the less be truly and really merited by metaphysics, even did the implicit truths, unconsciously embodied and latent in every form of science, never reveal themselves explicitly to any one for what they really are and what we are now recognizing them to be.

But in the last part of this essay its helpfulness will be much more fully revealed and made so evident that denial will be impossible save to those who either wilfully shut their mental eyes, or are unhappily unable even to open them.

One of the most important distinctions pointed out in the last part of this essay was the distinction between what is *objective* and what is *subjective;* and one of the most important facts therein put forward was the fact that, by the aid of memory, we can obtain entrance into the domain of objectivity, and know real existences external to our present feelings and imaginations.

Bearing this supremely important distinction and this fact in mind, let us advance further, and lay the fourth foundation-stone for the great temple of human science—beginning with the recognition of some very simple facts.

Physical science is emphatically experimental science. But every experiment, carefully performed, implies a most important truth, though it is so obvious a one that nobody would usually pay the slightest attention to it. Let us suppose that the experiment of cutting off a newt's leg has been performed in order to see whether it will grow again, and let us further suppose that it has grown again—as, in fact, it will grow again. This experiment will have demonstrated to us the certainty that such a thing is possible, because, in fact, it has actually occurred. But that very certainty implies a prior and much more important truth:

It implies the truth that if the newt has come to have four legs once more, it cannot at the very same time have still only three legs. This may appear to some of my readers to be altogether too trivial a remark. But there is nothing like a *concrete* example for making an *abstract* truth plain. Besides, it is almost impossible in such inquiries as these (metaphysical or philosophical inquiries) to be too careful in making each step as we go along plain, certain, and unquestionable. Therefore, it will not be superfluous here to distinctly note the fact that whatever it may be which we have become certain about because it has been proved to us by experiment, is only thus certain because we know that when a thing has been actually proved, it cannot at the same time remain unproven. If we reflect again on this proposition, we shall see that it depends on a still more fundamental truth which our reason recognizes—the truth, namely, that " nothing can at the same time both be and not be "—the truth known as "*the law of contradiction.*" This again, like the three certainties laid down in Part II., is a truth which carries with it its own evidence, and is incapable of proof. That such is the case a very little reflection will show. We constantly act upon it in daily life without adverting to it. The simplest rustic knows that if his wages have been paid to him they are no longer owing, and that if he has put his cart-horse in the stable it can no longer remain between the shafts. Yet the "law of contradiction" is but the summing up in an abstract form of a multitude of particular instances of this kind, as to each one of which no doubt is, or can be for a moment, seriously entertained by any sane mind.

If we were really to doubt about the law of contradiction we should thereby be landed in absolute scepticism, which we have seen to be mere folly, because all certainty would be thereby destroyed; for, if anything can at the same time both be and not be, then nothing can be *true* without its being possible for it also to be *untrue*, and this amounts to a veritable paralysis of the intellect, reducing us to mental impotency.

The excessive folly implied in any doubt of the objective validity of the law of contradiction has been so well shown by an attempt on the part of a Mr. E. T. Dixon to impugn its truth, that I think I cannot do better than here call attention to it. Mr. Dixon affirmed:* " If any one chooses to say a thing both ' is' and ' is not,' there is no law against his doing so, only if he does so he is not talking the Queen's English." To which I replied : " By so doing he breaks the law of reason, if not the law of the land ; and, indeed, to act on such a principle when on oath in a court of law might have inconvenient consequences." To a verbal quibble about the word " to be," I replied: " Let us avoid the use of the terms ' is ' and ' is not'; they are not necessary. Does Mr. Dixon really doubt whether if he had lost an eye he would still remain, after that loss, in the very same condition he was in before ? If any one does not see the objective im-

---

* See Correspondence in *Nature*, from December 10, 1891, to February 11, 1892.

possibility of such a thing *everywhere* and *everywhen*—i. e., if he does not apprehend the application of the law of contradiction—then he either does not understand the question, or his mental condition is pathological." The implications of science are really therein implied. Men may pretend to doubt them, their own existence, or the objectivity of mathematical truths. But their practice demonstrates their unfailing confidence in them on each occasion as it arises—as when cheated by false accounts, personally injured, or engaged in scientific research. When we enter the laboratory, we leave such follies outside. That nothing can simultaneously be existent and non-existent does not at all depend on the words employed to denote that truth, but is a law of "things." It would not lose its validity and objective truth, not only if there were no such things as "words" at all, but it would not lose it if the whole human race came to an end.

If any one who sees that the loss of an eye would modify his state of being here and now, does not also see that such a truth must also apply everywhere and everywhen—as before said—then his faculty of mental vision is one I do not envy.

For if we think of what the condition of things must have been a long time ago—in the days of Julius Cæsar, or when palæolithic implements were first fashioned—we shall see that the law of contradiction is as sure and certain with respect to the past as it is with respect to the present. We do not "*think*," we actually "*know*" with "*certainty*," that had Julius Cæsar been drowned off the coast of England he could not also have been assassinated by Brutus in the Roman Senate; and that if an early palæolithic flint man was holding a flint in his hand he could not at the very same time have had both his hands empty. The same certainty exists as to the most distant regions. We are quite sure that the moon's surface cannot be both irregular and absolutely smooth; and that the spectrum of a remote star, which shows certain definite lines, cannot, at the same time, be devoid of them. Such assertions may well seem utterly superfluous, yet the existence of men like Mr. E. T. Dixon shows that they are not so, but require to be distinctly noted.

When they have been noted and pondered over, then reflection reveals to us that this law of contradiction is not only implied by every physical science, and necessary to the validity of all our knowledge, but that it is an absolutely necessary and universal truth, which carries with it its own evidence.

But here, again, one or two of my readers may be startled by the words "absolutely necessary" and "universal." They may feel some vague doubt as to how this matter may be in the Dog-star now, or how it may have been long ages before our nebula was churned into worlds—supposing the solar system did so arise. I may be asked, "How is it possible that we men, mere insects, as it were, of a day, inhabiting a fleeting atom in an obscure corner of the universe, can know that anything is and must be absolutely true for all regions and the most distant ages?"

Yet, in fact, we know much more than

even this. We, poor and feeble creatures as we are, are endowed with power to see necessary limits to the action even of Omnipotence itself. For, let the reader first suppose that Omnipotence might have made our world such as it is, save that all coniferous trees were excluded from it. Then let the reader suppose that Omnipotence might have made our world such as it is, save that all its trees were conifers. The reader will then see that it is, and must eternally be, and have ever been, absolutely impossible even for Omnipotence to have made both these possible states of our world simultaneously actual. Having reflected on this simple but evident truth, he will be less disposed to doubt his powers of perception with respect to truths of a lower and more ordinary kind. It is necessary, indeed, to be careful not to declare anything to be certain till it has been seen to be clearly and indubitably true, but it is no less necessary that we should not shrink from declaring that to be true the certainty of which is evident to our minds, however wonderful it may be, and however inexplicable is the fact of our knowledge of it. We are able to explain how it is we know many things, and as time elapses we may come to explain our knowledge of very many more. But how we *know* primary and fundamental truths, which are self-evident and necessarily incapable of proof, must ever remain for us entirely inexplicable. Were they explicable, they could not be ultimate.

I think that this feeling of distrust as to our power of knowing absolute and universal truths is due to a second habit of mind which most persons have formed, and which runs parallel to the other mental habit which we before noted as having given rise to a prejudice against believing anything "on its own evidence." This second habit of mind has been formed as follows: Things which are very distant, or which happened a long time ago, are known to us only in roundabout ways, and we often feel more or less want of certainty with respect to them. On the other hand, we have a practical certainty about the circumstances and conditions which surround us at the present time. Thus we have come to associate a feeling of uncertainty with statements concerning things which are very remote. But nothing can well be more remote for us than the Dog-star, or time before the formation of the solar system. It is not then, after all, so surprising that this vague distrust should at first exist with respect to our knowledge of truths as being absolutely necessary and universal.

It is no doubt wonderful that we should be able to know any necessary and universal truths; but it will be seen to be less exceptionally wonderful if it be well considered in relation to our other active powers. It is wonderful: but then, deeply considered, so is every other atom of our knowledge. It is wonderful that surrounding bodies should be able to excite feelings within us—such as sensations of musical tones, sweetness, blueness, or what not. It is wonderful that, through sensations actual and remembered, we have perceptions of the various objects we from time to time perceive. It is wonderful that, on the occurrence of certain perceptions, we recognize our own existence past and present. So, also, it is wonderful that we recognize that what we know *is* cannot at the same time *not be*. The fact is so, and we perceive it to be so; we know things, and we know that we know them. But *how* we know them is a mystery indeed, and one about which it is, I think, perfectly idle to speculate. It is precisely parallel to the mystery of sensation. We feel things savory, or odorous, or brilliant, or melodious, as the case may be, and, with the aid of the scalpel and the microscope, we may investigate the material conditions of such sensations. But how such conditions can give rise to the feelings themselves is a mystery which defies our utmost efforts to penetrate. I make no pretension to be able to throw any light upon the problem " How is knowledge possible?" any more than on the problem " How is sensation possible?" or on the questions: " How is life possible?" or, " How is extension possible?" But, *Ignorantia modi non tollit certitudinem facti;*

and we know that we are living, that we feel, and that we do know something—if only that we know we doubt about the certainty of knowledge.

If we deny or doubt the "law of contradiction," we fall, as before said, into the unutterable absurdity of absolute scepticism, and we are thereby forced to admit that law's universal validity. But it is no mere law of our own minds, for if we are to listen to what our reason affirms to be evident, it is also a law which applies to things—to all things, from metals and other minerals to mental states. Such is the case, since we have seen it so declares with respect to the various instances we selected as examples. When we say that the number of oranges in a bag cannot at the same time be both "odd" and "even," we are certain that this is not a truth due to our organization, but to the real necessary conditions of existence of the oranges themselves. Our reason declares that the law of contradiction is no "form of thought" imposed on our intellect; but objectively certain, independent of our intellect.

To doubt this, then, would be to destroy the certainty of that which is the most evident to us of all propositions. It is thus a fundamental truth, upon which not only all reasoning depends, but which applies to everything which exists; since we see clearly that even a Supreme and Omnipotent Being could not—however different the existence of such a Being may be from our own—both exist and be non-existent, any more than such a being could cause one of the stars to be, simultaneously, both entire and divided into two separate halves.

Our perception, therefore, of the necessary validity of the law of contradiction teaches us both an absolute verity with respect to objective, external existences, as well as the existence of our own mental perception thereof: *e. g.*, that two things cannot be four things, and that we see such to be the case. This double perception (as to facts and thoughts) leads to the consideration of the last matter which it is here proposed to treat as fundamental. This matter is the answer to Pilate's question, "What is truth?"

It has been sometimes said that "truth is what each man troweth, and no more." But no rational man could seriously so affirm, since by so doing he would really refute himself—like the sceptic in declaring "nothing is true." For if truth were subjective only, that very condition would be a "fact," and every "fact" is something "objective" —more than an individual fancy. Therefore, if any one should say, "It is a fact that truth is merely an individual fancy," he would thereby affirm, "truth is merely an individual fancy and not merely an individual fancy," and so must explicitly refute himself. Putting aside such follies, it is evident that no man of science can reasonably doubt that truth is more than a mere quality recognized as belonging to a judgment by him who emits it. He must be sure it has a real relation to external things, or else his science could make no progress. We do not base scientific inductions and deductions on our knowledge of beliefs, but on our knowledge of facts; and, without a foundation of facts, beliefs are worthless. The truth of physical science consists in the agreement of "thought" with "things"; of the world of "beliefs" with the world of "external existences." Thus, if we state that "terrapins are toothless," we thereby affirm not only a correspondence between our subjective conceptions of "terrapins" and "toothlessness" and the objective realities—the real reptiles and their edentulous condition of jaw—but also a correspondence between our subjective judgment in making the statement and the objective coexistence of the "terrapins" and the condition we term "toothlessness." Truth, then, cannot be only "what each man troweth," but must be what a man troweth when he troweth in conformity with real external coexistences and sequences, and with the causes and conditions of the world about him. Truth, therefore, is, and must be, both subjective and objective. It is subjective regarded as a quality of any judgment of our own. It is objective as a quality of the judgments of any one else—and every one's judgments are objective save to the individual who so judges. The simple, but fundamental, truths to which attention has here been called—as being truths present, if unobserved, in the mind of every rational man—have most important and far-reaching consequences, and therefore merit the attention of every one who desires to be able to give a rational account of the convictions of his own mind. They are truths which are latent in every branch of physical science, and any *real* doubt about them would make not only experiment, but even observation, logically impossible.

And now we may advance to the first practical application of the various considerations which have been here urged, and may begin to build upon the foundations (of all our actual or possible knowledge) which we have here endeavored to lay—namely: (1) the fact of our knowledge of our own continuous existence and our successive states of consciousness revealed to us by memory; (2) the general principle termed the law of contradiction; and (3) the possibility of valid reasoning.

Knowledge is valuable for its own sake; the mind feels a natural and legitimate satisfaction in its acquisition as well as pleasure in its pursuit, and the pursuit of science may be its own abundantly great reward. Nevertheless, knowledge is for most of us largely esteemed as a means to guide us to useful and reasonable action, and our actions are certainly guided by our knowledge, as well as by our sentiments and feelings. Therefore, the more thorough and well-grounded our knowledge, the better must we be qualified for action; and this especially applies to our knowledge of ourselves. That knowledge, then, must be exceptionally helpful which best informs us respecting our own nature and our own powers. By it we may learn (1) not to waste efforts in directions which can lead to nothing; and (2) not to neglect actions which our nature demands we should perform, while supplying us with facilities for their performance.

But that we may learn what our nature really is we must first briefly consider that system of thought which forms the basis of agnosticism, and has become so widely diffused and popular. This is the system of *sensism, subjectivism*, or *phenomenalism* — to which I confidently oppose the system of *intellectualism.*

The whole scheme of human life—its powers, its aspirations, and its duties — is most powerfully influenced by the alternative to be here adopted. No help can be greater than such help as philosophy can afford us to attain to a clear perception of where the truth lies in this important controversy.

The sensist position may be thus stated:

(1) All our knowledge is merely relative.

(2) We know nothing with certainty save our own feelings.

(3) We can therefore have no knowledge of anything in itself—that is, as it exists objectively and independent of our knowledge of it.

The essential founder of this system was Hume. It was proclaimed by Sir William Hamilton, and has been rendered popular by John Stuart Mill, Alexander Bain, G. H. Lewes, Herbert Spencer, and Professor Huxley.

Some of my readers may ask, "How can

you venture to raise yourself up in opposition to a system which has such sponsors? To this I reply, there are two good reasons, one or both of which cause almost all of the above-named men to go wrong in such a matter; but I will deal with these later on, while a very small fact of recent history gives me a right thus to raise myself in opposition.

But first I would beg such questioners to note that philosophy is not a matter of *authority*, but of *reason*. At the beginning of this essay I made my appeal to the reason of my readers, and I now repeat that the authority of no name, however justly or unjustly respected, can have the least weight in the balance of science against the clear declarations of the individual intellect.

The very small fact of recent history which gives me some claim to be heard in this controversy is the fact that, when my intellectual life began, I was a student and disciple of the very school I here oppose. The works of Hume, Hamilton, Mill, Lewes, Bain, Spencer, and Huxley I studied to the best of my ability, while I have had the great advantage of personal acquaintance with some of the more distinguished ones who were my contemporaries. This has given me a much better grasp of the system they favored than could have been obtained by reading only.

Absurd as that system is, it for a long time held my mind in thrall; but if such considerations as I have here put forward had been brought to my notice in my earlier days, I should have been spared much waste of energy and much disadvantage, and I therefore feel keenly the evils of such a system. As it was, I after a time began to feel doubt about the validity of the system I had, at first, ingenuously adopted.

It took me, however, a long time to find my way out through the ingeniously constructed metaphysical labyrinth in which I had got myself imprisoned. But I escaped the philosophic Minotaur (agnosticism) not by drawing back or shirking any difficulty, but by pushing forward and slowly working my way through its difficulties till I made my exit, after having thoroughly traversed it. But to return to the system itself:—

In the second part of this essay it has, I hope, been satisfactorily demonstrated that we can know our own continuous existence, and, by memory, obtain a knowledge of things objective. If so, the second and third of the above cited sensist positions *ipso facto* fall to the ground. But over and above memory, a comprehension of the law of contradiction also introduces us into the domain of objectivity—of real things in themselves, as distinguished from thoughts and feelings.

As to the assertion that "all our knowledge is merely relative," there is a noteworthy ambiguity. Of course, anything which is "*known to us*" cannot at the same time be "*unknown to us*"; and, so far as this, our knowledge may be said to affect the thing we know. A thing known must stand in the relation of being known, and our knowledge is "relative," inasmuch as it could not be knowledge without that relation. But this is utterly trivial. Our "knowing" or "not knowing" any object is—apart from some act which may be the result of such knowledge—a mere accident of that body's existence, which is not otherwise affected thereby. If the fossil remains of three porpoises are discovered in some rock, were they not as really there before they were known to exist, and would they not have been there if they had never been discovered at all? As I pointed out when considering our knowledge of our own existence, nothing exists by itself and unrelated to any other thing; but that does not make our knowledge relative, save in the above mentioned trivial sense.

But, indeed, the system of the "relativity of knowledge" really refutes itself; for every system of knowledge must start with the assumption, implied or expressed, that something is true and known so to be. By the teachers of the doctrine of the "relativity of knowledge" it is taught that the doctrine of the relativity of knowledge is thus true. But if we cannot know that anything corresponds with external reality, if *nothing* we can assert has more than a relative or phenomenal value, then this character must also appertain to the doctrine of the "relativity of knowledge" itself.

Either, then, this system of philosophy is merely relative or phenomenal, and cannot be known to be true, or else it is absolutely true and can be known so to be. But it must be merely relative and phenomenal, if everything known by man is such. Its value, then, can be only relative and phenomenal; therefore it cannot be known to correspond with external reality, and cannot be asserted to be true. If anybody asserts that we can know the system of the relativity of knowledge to be true, he thereby asserts that it is false to say that our knowledge is only relative. In asserting that we can know the system of the relativity of knowledge to be true, he who so asserts thereby proclaims that some of our knowledge must be absolute; but this upsets the foundation of the whole system.

Therefore, any one who upholds such a system as this may be compared to a man seated high up on the branch of a tree which he is engaged in sawing across where it springs from the tree's trunk. The position and occupation taken up by such a man (whatever his name or repute may be) cannot be considered as evidence of his possessing any exceptional amount of wisdom.

If, then, the system of the "relativity of knowledge," or "sensism," be false, and the dictates of our reason when probed to its ultimate foundations are to be trusted—*i. e.*, if we avoid the mentally diseased condition of absolute scepticism—we have three clear intellectual perceptions or intuitions: (1) a perception or intuition of intelligence, or intellectual activity, by a consciousness of it as existing in ourselves—in our own intellect; (2) a perception or intuition of things external to us and of extension in the extended objects about us; and (3) a perception or intuition of a divergence and distinction of nature between thought and extension—between these two very different kinds of existence. We can perceive in our own very self both a continuous thinking intelligence and extended structures (such as, *e.g.*, our hands and feet) which do not think, as we know by our own consciousness; but a little more reflection on what our own consciousness tells us about ourselves teaches us a further more important lesson. For we know most intimately, by and in our own consciousness, something—our own intellect —which, as we have seen, exists continuously, which is conscious of successive objects and events, and which, while itself transcending them, can recognize them as forming a series which it can contemplate as a whole or in parts, and in different orders as may be desired. In other words, our mind knows some of its thoughts and experiences of yesterday, and also that itself is something beyond them. It can recognize those thoughts and experiences as having occurred in a definite succession, and can think of them as one whole ("the thoughts and experiences of yesterday"), or of a part of them, and can think of them in the order in which they occurred, or the reverse order, or in any order which our mind may for any reason select. This intellectual power or principle (the mind) also knows itself with perfect certainty (as we have seen), is aware of the kinds and directions of its activities, and can regard them as a whole, or in groups, or singly. It can, it well knows, perceive its own states both passive and active, and also external objects and events, recognize what it thinks they are, and can compare the relations between them, returning upon itself at will along different lines of thought, and also setting forth in various directions at will. Such a power, or principle, aware of all these things and present to them all, cannot itself be multitudinous, but must be a unity as complete as it is possible for us to imagine—namely, a simple unity. Therefore, this principle can be neither a material substance nor a physical force, but presents the greatest contrast to both; and the greatest contrast to what is material is what is immaterial. Also, since each man knows that it is he who not only thinks but also feels, he also knows, if he reflects upon it, that his body and this thinking principle are one unity.

Therefore, each of us is a unity with two

sets of faculties: material and mechanical on the one hand, immaterial and unmechanical on the other. No certainty we can attain to about any other object can be nearly so certain as is this certainty we have about our own being; above all, its dynamic or immaterial and active aspects. That each man is a material, definitely organized substance in one unity, with an active, immaterial principle revealed in consciousness, is really the first truth of physical science. It is emphatically the fundamental truth of the science of living organisms (biology), for of no other living thing can we possibly have so complete a knowledge as of ourselves, since only to self-scrutiny can that, the most direct kind of knowledge—our immediate consciousness—be directly and immediately applied.

I have made the preceding observations here because I desire to call special attention to the effect which the study of our own intellectual activity must have on any comprehension of the world about us and our relations thereto. In the historical order of knowledge—the knowledge of the individual as well as of the race—the study of what is external to us does come (as before said), and must come, first. But as we emerge from and put away the things of childhood we have more and more to reverse this process. For any satisfactory comprehension of the world and man, the first and most pressing need is an adequate acquaintance with our own higher mental powers and processes. The certain, though gradual, effect of their faithful study will, I am confident, be the overthrow of "*sensism*," and induce its replacement by the only system which is satisfactory in its application on all sides, the system which distinctly recognizes that the basis of all science must consist of truths recognized by thought as self-evident and necessary—the system of "*intellectualism*."

I most earnestly desire to direct my readers' attention to the antithesis which exists between these two schools of thought and to the consequences which either involves.

The intellectual world is rapidly dividing into two camps: (1) those of one upholding the validity of the fundamental dictates of reason with all its consequences; (2) those of the other who refuse to admit the validity of its dictates, and are forced in consequence to stultify themselves by refuting their own system and denying all absolute and objective knowledge, even their perception of their own continuous existence.

It is then, as before said, my most earnest desire, by appealing to the fundamental declarations of each man's consciousness, to make evident the nobility of human reason by showing its wonderful power of knowing objective as well as subjective truth, and of recognizing the actual certainty of that which it sees to be evidently true. This is so supremely important, because, if we could not trust its declarations as to truth, how could we confide in its imperative behests as to goodness? And here we touch the foundation of all that is noblest in human life, and the final guide of aspiration and of conduct.

The greatest evil of such systems of philosophy as those here combated is that, by insidiously undermining our rational confidence in human reason, they weaken the springs of vigorous and healthy action. Such systems may well be called "the art of losing one's way methodically."

The indignation of any one who values our highest mental powers may well be excited by sophists who make use of exceptional mental gifts for the purpose of disparaging and virtually denying the existence of that wonderful and admirable human intellect which they insult and blaspheme.

The "helpful science," then, is not only the real, though generally unsuspected, source of every scientific discovery and all artistic progress, but also helps us, in a way in which nothing else could help us, to appreciate the dignity of our own nature. We have next to consider, by its aid, what that dignity involves.

An inevitable instinct impels us towards seeking our own happiness and gratifying our desires and passions. But however we may allow ourselves to be carried along, our reason can nevertheless ask the question, "Are we really acting in a reasonable way if we make pleasure our deliberate aim in life?"

Now, evidently the only rational aim of life is that which reason tells us *should* be its end. But to say that anything "should" be its end is equivalent to saying it ought to be its end, and the word "ought" is meaningless apart from the conception of "duty." Thus our *reason* tells us (however we may *feel* about it) that "duty" and not "pleasure" is the true and proper end we should pursue.

I have said "however we may *feel* about it," because it is all important to bear in mind the profound difference which exists between the "feelings" and the "intellectual perceptions" which may accompany our actions.

Doing things which are right and kind is often accompanied by pleasurable feelings, but such things may be very painful.

Let us suppose that a man has found out that some property which has enabled him to support a large family is not really his property, but belongs to some one else, to whom he is bound to hand it over. Let us further suppose that he does so hand it over, because he sees clearly that it is his duty so to do. His perception as to what is right will not prevent his feeling the material disadvantages which attend his praiseworthy action or change pain into pleasure.

Similarly, a man or woman may judge that it is his, or her, duty to break off an intercourse which is against conscience, without its being a bit less distressing, even heart-rending, to break it off.

It is plain that we may feel pleasure in doing things which are wrong, for certainly they would otherwise never be done. On the other hand, much painful regret may be felt on account of quite innocent actions. Thus some trifling breach of etiquette, or some harmless violation of social usage may cause us to blush and feel far more shame than might attend some considerable moral delinquency. Keen remorse also may be felt on account of having neglected some excellent opportunity of pushing our fortune, or even of committing some very pleasurable but very immoral action. A French writer has said that no regret is so keen as the regret which may be felt for the non-commission of pleasant sins which might have been enjoyed.

Charles Darwin has said: "Conscience is that feeling of regretful dissatisfaction which is induced in a man who looks back and judges a past action with disapproval." Now, "conscience" certainly "looks back and judges," but not all that "looks back and judges with regretful dissatisfaction" is conscience. Otherwise, a libertine who, wishing to enjoy an exceptionally immoral performance, found to his surprise that he—having mistaken his theatre—was witnessing an innocent one, might exercise his "conscience" in looking back and judging "with regretful disapproval" that he had *bought a wrong ticket!*

"Conscience" is a judgment of a particular kind—namely, about "right" and "wrong" —and nothing else.

The man we have supposed to be conscious that he had made some trifling slip in his manners while in company may color up and feel distressed and ashamed, but he will not on that account judge that he has committed an action *morally wrong.*

The profound distinction which exists between our ideal of "goodness" and every other conception can be illustrated by any—even the most trivial—statement concerning duty. Let us suppose, for example, that any one is told he should "pay his tailor," and that the assertion is disputed; how should we set about trying to convince him of its truth? Evidently we should have to fall back upon some more obvious and general assertion about duty, such as the assertion, "Every man is bound to pay his debts."

If this is again disputed, we should have to urge some still wider precept, such as "A man is bound to carry out an agreement he has himself chosen to make," and so on.

Every step we take to explain *why* any duty should be performed must consist of some still more simple assertion of the kind, till we come to an assertion about duty, the truth of which is admitted to be self-evident.

Now, all our certain knowledge must be either evident in itself, or must depend upon some other knowledge which is evident in itself. As we have before seen, we cannot go on arguing forever, and every proof must stop somewhere — namely, when we reach what is evident without proof.

If, then, we want to urge some statement about any particular action being "right" or "wrong": if that statement be not admitted to be evidently true, we can only prove it to be so by means of some more general and elementary statement of the same nature. Therefore, *the judgments which lie at the root of any system of thought about ethics* (about right and wrong) *must themselves be ethical.*

This truth cuts the ground from under—renders simply impossible—the theory that any judgment as to moral obligation could ever have grown out of mere feelings of liking or disliking, sympathy or aversion, goodwill or hostility of our fellow-men.

No stream can possibly rise higher than its source. "Social approbation," then, could never have produced the conception of "right and wrong"; for how could a mere habit of obeying society have ever led a moral hero to denounce that habit and defy society?

Yet there are a number of men—such as Mill, Bain, Spencer, Huxley, etc. — who, while they lay down the law as to what "ought" or "ought not" to be done, deny that there is any real, fundamental distinctness between "virtue" and "pleasure." I mean they deny that there is any absolute distinction between them, and affirm that "good actions" are merely actions pleasurable or useful to the individual who performs them, or advantageous to his fellow-men. They say, also, that it is the pleasurable or useful *results* which cause actions to be good actions, not the intentions with which the doer may perform them.

It is true, we say, "That is a 'good' knife," because it cuts well, and any weapon, coat, or other useful article is said to be a "good" one if it serve well the purpose for which it was intended. But a very little consideration will show that such a use of the word does not bring us to the fundamental meaning of the term. For "conformity to an end " will not make an action good unless the end aimed at is itself good and agreeable to duty—unless by conforming to it we "follow the right order." If a youth, carefully instructed by Fagin, conforms to his end by picking pockets with extraordinary deftness, such " conformity " will not make pocket-picking "good."

But when the end is really good and part of our duty, if we ask, " Why should we do our duty? why should we follow the right order?" there is no answer possible but, "It is right so to do."

If it be urged in opposition that "we should follow the right order because it is our true interest," he who so urges must either mean "we should always follow our own interest," which is equivalent to abandoning the rule of " right and wrong " altogether, or he must mean " we should follow our interest, not because it is our interest, but because it is right "—a proposition which is a mistaken one, but yet one which, in its mistaken way, affirms the very rule of "right and wrong" it was designed to oppose.

But persons who say that the morality of any action depends on its results can always be refuted simply by examining into the assertions about duty which they themselves make. One most amusing, instructive, and memorable example of the kind is afforded us by no less a person than John Stuart Mill. That eminent utilitarian (*i.e.* denier of the absolutely distinct natures of "goodness" and "pleasure"), has written * as follows:

"If I am informed that the world is ruled by a being whose attributes are infinite, but what they are we cannot learn, nor what are the principles of his government, except that the highest human morality which we are capable of conceiving does not sanction them; convince me of it and I will bear my fate as I may. But when I am told that I must believe this, and at the same time call this being by the names which express and affirm the highest human morality, I say in plain terms that I will not. Whatever power such

---

* See his *Examination of Sir William Hamilton's Philosophy*, p. 103.

a being may have over me, there is one thing which he shall not do—he shall not compel me to worship him. I will call no being 'good' who is not what I mean when I apply that epithet to my fellow-creatures; and if such a being can sentence me to hell for not so calling him, to hell I will go."

Now this declaration is truly admirable; nevertheless, it is very much out of place in Stuart Mill's mouth, or in the mouth of any other professed "*utilitarian.*" For, if actions *were* "good" or "bad" *merely* according to the pleasure or pain which followed them, and if by flattering a bad God we could get the greatest happiness, while by refusing to do so we should incur the greatest misery, then clearly on *utilitarian principles* (not on those. I advocate), such flattery would be "good."

Stuart Mill's position is indeed a curious one; for, of course, he means that other men should do as he says he would do. Thus on the one hand he declares that all men should (as being utilitarians) seek the greatest happiness for all, and also that in so doing they should all voluntarily plunge into the greatest misery!

This is one striking example of how impossible it is for the most distinguished men to go directly in opposition to the principles of the "helpful science" without stultifying themselves.

And now as to Professor Huxley's assertions (1) that it is not the *intention* of the man who acts, but the *consequences* of his act which makes his action "bad" or "good," and (2) that the highest virtue is to do good without thinking about it:

Let us suppose that two men have each a sick wife, and that the doctor has left with each man two bottles: one a valuable internal medicine, the other a poisonous lotion to be used externally. One of these men, who loves his wife fondly, gives her, by pure mistake, the lotion to drink and kills her. The other man desires to poison his wife, but, by a similar mistake as to the bottles, gives her, unintentionally, the right medicine and cures her. Can there be any doubt as to who is the truly guilty man? Who dare assert that the action of the second man was a "good" action because, in spite of his evil intention, it had a good result?

Yet this must be so if Professor Huxley's principle has any truth in it!

Then as to "*doing good without thinking about it*"—it cannot be meant that it is the absence of thought which makes a spontaneously performed useful action specially meritorious; otherwise we should attain the climax of virtue by performing beneficial actions unconsciously in a state of somnambulism!

The admirable nature of spontaneous good actions lies in their being the result of good habits, not in the mere absence of reflection. A man cannot appreciate *justice* without being able to distinguish it from injustice, and to love "goodness" he must at least "know" it.

But another objection to any absolute distinction between "right" and "wrong" is sometimes drawn from the fact that different nations (and the same nation at different times) take different views as to the "goodness" of some particular kind of action. But this argument is quite valueless. It would be absurd, indeed, to suppose that all men were naturally, or supernaturally, furnished with a whole code of laws directing what is to be done and what abstained from in all cases. What is affirmed by Intellectualists is, that men are naturally and universally (idiots apart) endowed with a perception that there *is* such a thing as "right" and "wrong." Men are not necessarily devoid of morality because they draw their lines and rules in different places from what we do. The most horrible actions, such as the deliberate killing of aged parents, may really be the result of true moral judgments under peculiar circumstances. Such is the case in a tribe the elders of which desire death to secure a happy immortality. Men do not always agree about the *application* of moral principles: what they agree about is that there is such a thing as a moral character attaching to certain actions. Even an untutored savage would perceive that an ungrateful and treacherous injury inflicted on himself was a wrong, and one that merited chastisement.

Though thieving may have been here and there inculcated, there is also the saying "Honor amongst thieves," and the greatest rascals often recognize the moral claims their "pals" have on them. Men have often thought it right to do things which were really unjust; but they never thought any action to be "right" *because* it was "unjust," or any action to be wrong *because* it was "just."

On the other hand, we may very properly deem some actions to be less virtuous—less "right"—than others *because* they are useful; and others less "wrong" *because* they are useless. Thus we may admire the devotion with which a man may minister to a sick friend; but if, after that friend's death, we find that the man who ministered to him has received a rich bequest, our appreciation of his devotedness may be much diminished. Similarly, a woman may have a sinful attachment to a man, but if we find that instead of any worldly gain thence derived, she simply sacrifices herself for him, our censure may thereby be mitigated, since it shows she "has loved much."

Thus so essential is the distinction between "the good" and "the useful" that not only does the idea of "benefit" not enter into the idea of "duty," but the very fact of an action not being beneficial may make it more praiseworthy. Its merit is increased by any self-denial which may be necessary to its performance, while gain tends to diminish the merit of an action. It is not that absence of gain or pleasure benefits our neighbor more; it is that any diminution of pleasure which circumstances may occasion (irrespective of any advantage thereby occasioned to our neighbor) *in itself* heightens the value of the action. But evidently that can never be the "substance of duty" which increases "dutifulness" by its absence!

We have seen, in considering the principle of contradiction, that the human mind has the power of knowing absolute, necessary, and universal truth. Similarly, reason shows us that the commands of conscience are absolute and supreme. But if we see that its universal and unconditional authority can never be evaded, then the ethical principle must be rooted, as it were, within the inmost heart, in the very foundations, so to speak, of the entire universe, which it must pervade wherever any moral beings exist. The principles of the moral law must then be at least as extensive and enduring as are those starry heavens which shared with it the profound reverence of Kant.

The conception of duty is the conception of something supreme and absolutely incumbent upon us without appeal—apart from any question of pleasures and pains, rewards and punishments. It is, in the immortal words of Cicero:

"*Quod tale est ut detracta omni utilitate sive allis præmiis fructibusque per seipsum possit jure laudari.*"

Part IV

THE conclusion we arrived at in the third part of this essay was that the ideas "virtue" and "utility" are fundamentally distinct, and also that the claims of "duty" are absolutely imperative and supreme. This conclusion, it was observed, had certain important antecedents, and it has also no less practical consequences.

The antecedents I referred to were the considerations which warranted that conclusion. In order to obtain a just appreciation of these it was necessary to dig down to the very foundations of human knowledge and so ascertain what must be the ultimate grounds of our beliefs in every proposition which we regard as the most certain and unquestionable.

It is impossible, we have seen, to find, or even conceive of, grounds more absolutely certain than those which ultimate truths bear within themselves—namely, their own self-evidence. The ultimate truths we have recognized are (1) our own existence and our knowledge thereof; (2) the validity of our own perception of objectivity through our faculty of memory; and (3) the validity also of logical reasoning; the certainty (4) of the principle of contradiction, and (5) the reality and significance of "truth." By considering the wondrous fact that the human mind can know a certain amount of universal and necessary truth (as that nothing can simultaneously exist and be non-existent), perceive the extension of extended objects and the difference between "thought" and "extension," we recognized that our intellectual energy must be an immaterial unity and one faculty of our whole being, the material nature of which possessed the other property—namely, "extension."

These truths follow one from another as necessary consequences of our fundamental perceptions of things evidently true, and are the antecedents and justification of our ethical perceptions—our judgments of what is certain in the domain of morals.

Let us next turn to some of the consequences of our moral intuitions.

Every man who has pondered over his own nature (without having accepted as satisfactory any of the current forms of religious teaching) must have speculated about his origin and his destiny, and have probably asked himself, "What is the meaning of life?"

Every such man has his own way of regarding the world, and, it may be, has arrived at the conclusion that he can attain no satisfaction about it. He will, however, certainly admit that, to obtain answers to such questions—if they can be obtained—is a most practical matter. To such men this essay is offered as a presentation of thoughts (a science) to them specially helpful. If the rational end of life is (as put forward in Part III.) the fulfilment of duty, then no one can have an objectless existence. The poor no less than the rich, the foolish as well as the wise, and those who are aged and infirm, as well as the healthy and young, have all a truly noble prospect before them. The fulfilment of duty has also, as I think may be clearly shown, its practically advantageous results.

I observed, at the conclusion of the last part of this essay, that reason shows us the fulfilment of duty to be incumbent upon us apart from any question of prospective pleasures or pains. But, though duties are thus incumbent apart from questions of rewards or punishments, yet we certainly have evident perceptions as to justice with respect to the consequences of our actions.

It is an evident truth that every responsible person deserves differently according to the way he has acted with respect to his responsibilities. If he has acted well, it is plain that he has so far acquired merit, and has claims in accordance therewith. If he has acted ill, it is manifest that he has acquired demerit, is less to be valued, and deserves to be treated accordingly.

Justice demands that each man should be dealt with exactly according to his deserts, and that happiness should (at least, ultimately) attend on virtue, and unhappiness on vice. Not only does the practice of virtue often entail grave sacrifices, but, as we have seen, the greater the sacrifice the higher the virtue and the more evident the merit. Although a virtuous man feels, as a rule, pleasure in his good actions, yet to this there are limits, and it is surely desirable that evil should be avoided and good deeds done by the great mass of mankind, who certainly have not attained a degree of virtue which any one need term Quixotic. It would outrage our perception of what is just, did we know that happiness would ever be finally divorced from virtue; and this perception is one which forcibly shows the close relation which exists between "knowing" and "doing."

What are the plain dictates of ordinary common-sense? what maxim do we hear urged in the by-ways of our cities which comes home not only to the feelings, but to the moral judgment of the masses? Surely it is the saying that a man deserves "a fair day's wages for a really fair day's work."

And as every man, whatever his degree, needs his *daily bread*, so also is a *daily work* required of him, as his conscience plainly tells him. The daily work required of every one of us is to "do what is right"—to "follow the right order"—according to our various circumstances: and this in our secret thoughts as well as in word and deed; for thoughts deliberately entertained are actions begun. This is the work set before every man and woman to do—rich and poor, strong and weak, young and old, learned and ignorant —and this is the supremely "good work" which our perceptions of justice assure us cannot ultimately be divorced from happiness. Yet both ancient and modern writings teem with protests against the successful evil-doer, and history is full of examples of apparently successful injustice. It is true that some agnostics affirm that every man receives microscopic justice—complete retribution—in this life for his every deed, word, and thought. I recollect, a few years ago, appealing to a leading agnostic in England for a subscription towards maintaining the widow and children of a poor but most worthy laborer, suddenly killed by a fall from a house-roof. He had left behind him, bed-ridden with cancer, a wife to whom, and to his young family, he had been accustomed to minister before setting out for his daily toil. I did not obtain one shilling from this leading agnostic. He did not use the precise words "Served him right" (!), but he said to me, "How do you know he was not entertaining a bad thought when he so fell, and met with his punishment?"

So groundless a superstition as that every one has such microscopic justice awarded to him here, hardly seems to me to deserve argument; but is it possible to think that the poor boy who was the son of Marie Antoinette, and who suffered so many cruel tortures, could by the actions of his brief life have merited so terrible a chastisement?

If, then, we cannot count on justice here, our ethical perceptions seem to demand a future life as a moral necessity.

It surely must be admitted that a strong conviction as to a future life is of the utmost practical importance with respect to the moral work we have some of us every now and then to do. If we can attain a certain and sure knowledge that such a future is even probable, such a *knowing* has evidently a very close and influential relation to *doing*. I am far indeed from meaning to imply that we should do good for the *sake of a reward*, but, knowing what sort of creatures we men are, I cannot shut my eyes to the fact that for most of us—when the will, under temptation, is just trembling in the balance—such considerations must have, and for reasonable men should have, a decided influence.

As to the reasonableness of any confidence in a future after death, as considered with the best aid we can obtain from "the helpful science," I have something to say later on. Here, however, we may at once note what considerations as to physiology may be able to teach us. But, in fact, the very last refinements of physical science do not add one jot to the most commonplace and old-fashioned arguments against a future life. It was known centuries enough ago that "when the brains were out the man would die, and then an end," as regards the life of the body, and we know absolutely no more to-day. We are, of course, utterly unable to *imagine* the soul's existence after death, because we have had no such experience, and we can never imagine anything whereof we have had no kind of experience.

But our inability to form a mental picture of anything is no reason whatever for not believing in its existence. Who can picture an act of thinking?—yet nothing is more certain to us than that we not only believe, but absolutely know, with entire certainty, that we can and actually do think. Similarly, our intuition of the self-evident character of the precept "We should act rightly," our responsibility with respect to our actions which consciousness reveals, and the manifest absence of any complete system of retribution in this life, positively demand a life hereafter. The demand will, when we consider what has been already pointed out about the varied powers of our simple intellectual unity of nature, appear a very reasonable demand.

Either the moral law is a fable, and Jack the Ripper a utilitarian who blamelessly follows after pleasure in his own peculiar way, and the grossest and most malicious of mankind are alone right and alone rewarded, or virtue must sooner or later be vindicated and victorious.

But these reasonings may be combated by some such objections as the following:

"We can know nothing about such matters; for a future life in accordance with our deserts implies a God, and as to the existence of a God we can know nothing. Eminent agnostics assure us that, ignorant as we are of all such matters, we are above all incapable of knowing anything about a cause of the universe. We are, indeed, rather proud than otherwise of our ignorance, since we are convinced that you, in thinking you know something even about a God, prove thereby that you know less than we do ourselves. We altogether deny that any knowledge of a cause is possible at all; what is called 'causation' is really nothing but invisible, unconditional 'sequence.' We can see that one thing, or set of circumstances, is always followed by another, but we never do or can see any bond or compelling nexus between them!"

In reply to these objections a few words must be said about our knowledge of causes of any kind, before entering upon the tremendous question concerning the existence and nature of a cause of the whole universe, and as to the possible extent of our knowledge thereof.

It is quite true that we never feel or see physical causation, because it is both intangible and invisible; but, though our senses cannot perceive it, our intellect can and does. When we knock a nail into a board with a hammer, it is simply nonsense to tell us that, because we can only see the nail, board, and hammer, we cannot *know* that we exert a force which *makes* the nail go in. But there is one instance in which a man can be aware not only of an antecedent and consequent, and the causal relation between them, but also the very bond or nexus between them can be distinctly perceived by us. This is the case whenever a man is in doubt about what course to pursue, owing to his being drawn in different directions by different motives. Then the inflow and force of the conflicting motives acting upon his own mind can be distinctly perceived by him. We can all also perceive it when anything resists our will. Let us suppose that the stem of a small tree has been partly sawn through and that we then try whether we can pull it down. If the coherence of the part not sawn through is still very great, we may have to exert all our force to overcome it. When at last we have succeeded, and are exhausted with our efforts, we may feel very vividly that any one who denied we had *caused* the tree to come down must be as great a lunatic as any one who denied the objective existence of the tree at all. In fact, the idea "force" is a primary and ultimate idea which cannot be analyzed into any others. If any reader doubts this, I can only advise him to try to analyze it.

But it may be said—because such follies have been printed—that, though we may be conscious of our own force, we err if we assert efficient causation in any other instance, making the mistake even of attributing to inanimate things feelings like those we experience in making physical efforts. Surely greater nonsense has been rarely uttered. Let us suppose the partly-sawn-through tree to be not even touched by us, but that a gale has sprung up which, after having swayed it to and fro, breaks it off and prostrates it just as we have supposed it prostrated by human efforts.

Are we not to say that the wind has exerted as much force as was ours? and can we not say this confidently, without being such idiots as to attribute " feelings" to the wind?

Truly, then, we have actual experience of causation, but we have much more than that, for a little reflection will be enough to show the reader that the law of causation is a necessary and universal truth which carries with it its own evidence.

In the first part of this essay (among other illustrations of the certainties of every-day life) we said that if we find a door closed which we knew was open some time before, we may be certain that some person or thing has closed it. Similarly, if we come upon a corpse with its throat cut, we know with certainty the wound must have been either self-inflicted or been the act of some other person. All certainties of that kind are summed up in the law of causation, which declares that " Every change is due to some cause." This being a primary, ultimate truth, cannot, of course, be proved by any other truth, yet its certainty may be perceived by reflection as well as directly. Thus it is plain that what does not even exist cannot act, and, consequently, cannot be a cause. Therefore, anything which comes newly into being cannot be caused by itself, because it could not possibly have acted before it came into being. It must, then, have been brought into being by the agency of something else which was its cause. Every change in anything which already exists is, in fact, a new mode of being, and, therefore, equally demands a cause for its existence. It must, then, be due either to some distinct existence or to some other antecedent mode of being of that thing which now exists in its new mode. As, when we awake from sleep, our awakening must be due either to something external which has awakened us, or to some change which has taken place in our own organism. In the latter case, that change or new mode of our being, which we call "wakening from sleep," had for its cause an antecedent state of our body—more vigorous circulation, or what not.

Again, all and every object we see or feel must, we know, be the result of the action of some or other external cause or causes. This is evident with every object made by man; but no stone we tread on, no patch of sod or mud, can have come to be what it is save by the action of some antecedent causes. And this does not only apply to every complex structure, but to the simplest material object, even though it consists alone of what we call one of the ultimate chemical elements; *e. g.*, a pure metal or piece of crystallized carbon — that is, a diamond.

Any such object demands a cause for its being in the place it is at the time it is there, for its size, shape, etc., and for all its relations to surrounding things, as well as any special qualities of its own internal condition. Its own special conditions would also demand a cause, even if such a body existed alone and by itself in an otherwise empty universe—if we can permit ourselves to frame for a moment so absurd an hypothesis. Therefore, everything which can be seen not to contain a sufficient cause for its own existence within itself, must be due to some cause or causes external to it. Only something which is absolutely simple, indivisible, and eternal can escape from this law of universal causation. Moreover, this perception is not the mere result of a mental impotence of the imagination—it is not a negative inability to imagine a complex thing uncaused—but is a positive and active power of perception. Let the reader first consider his idea of a stone of some definite shape and size, made of two or more mineral substances. Then let him ask himself whether he does not actively and positively *see* that its shape and composition must positively be due to influences of different kinds, or whether he finds himself merely passive, and unable to help himself to be certain of anything about it.

This consideration does away with an objection made to the law of causation when, as is often the case, it is incorrectly stated. It is sometimes stated thus: "Every existence must have a cause." Then, of course, the objection at once arises that thus we have a *regressus ad infinitum*, and, if there is a God, he must have a cause, and so on forever—which is, of course, absurd. It is not, however, "everything which exists," but "everything which *newly* exists and does not contain within itself a sufficient cause for its being," which alone demands a cause.

Having gained a clear perception of this self-evident and necessary truth, let us consider the material universe in the light thus obtained.

The universe is the theatre of incessant and apparently unending change; but our present object is not to examine any of those changes or any number of them, but to consider them as forming one great, unimaginably complex whole. Does science unequivocally point to any beginning of such a whole? I do not see that reason makes it evident either that the whole cosmos (considered as one vast unity) ever had, or had not, a beginning, or will ever have, or not, an end. It is, at least, conceivable that the cosmos may be a real system of perpetual motion.

Of course, if the whole universe ever had a beginning, it must have had a cause; but how is it on the hypothesis that it never had a beginning? We saw that even a body composed of a single chemical element demands a cause to explain its shape and internal conditions. *A fortiori*, a body composed of many substances, with a complex internal structure and a variety of internal energies actively at work, demands a cause, and not one bit the less if it exists eternally.

This must apply equally if we imagine this body expanded indefinitely till it reaches the extent of the material universe. If such a universe is eternal, it none the less necessarily demands an eternal cause. This is absolutely certain, because the universe, as one whole, could never have been evolved by a process of natural selection — that is, have proved itself able to survive in a competition with others—because the universe, considered as one whole, could have had no competitors, and so, if it existed eternally, it must have eternally existed by itself, with the exception of its cause.

What, then, must we say as to the nature and attributes of such a first cause? The universe known to us is a universe replete with order, harmony, and beauty, and is governed by laws admirably correlated. It is also the abode of at least one race of beings (ourselves) who possess intellect, can apprehend truth, goodness, and beauty, and possess a power of will. Moreover, we human beings did somehow come into existence in a world (the earth) previously devoid of organisms endowed with such marvellous faculties.

Now our reason assures us that we can, to a certain extent, judge of causes by their effects. We may be quite sure that a donkey-engine will never draw a heavy train from New York to San Francisco; and whatever may be produced, must have been produced by something adequate to produce it—and this applies to a first cause as to every other cause.

We too are, however, sometimes told that we can be no judges of the adequacy or inadequacy of causes, and we are asked how we could know—*a priori*—the "adequacy" of a piece of steel to produce a wound, or of a flame to produce a burn? To these objections it may be replied that the "adequacy" is not in the steel or in the flame, but in these as affecting a sensitive organism which they may injure. The organism and the agents are together adequate to produce the effects cited, and the reason can perceive their adequacy.

But the one appeal of physical science is to "experience," and what does experience tell us? We have certainly no experience of life having been produced from the lifeless, or of sensibility and intellect appearing without their pre-existence in the agents which caused their existence. In short, experience shows us that "*Nemo dat quod non habet.*" Now, among the effects produced quite recently (geologically speaking) in this planet, there are three which are notably distinct from all the others. These are: (1) intellect which can perceive truth; (2) intellect perceiving the distinction between right and wrong; and (3) the power of will.

Therefore, a first cause, as adequate to produce such effects as these, must possess corresponding attributes—namely, intellect, goodness, and will. Of such attributes our own intellect, goodness, and will, can be at most, as it were, but faint reflections. In other words the First Cause, as possessed of intellect and will, must be personal—that is, a Personal God. By the term "personal," nothing more is here signified than the possession of these two attributes: there can be no other resemblance to a human personality.

Intellectualists are often reproached with what is called their "Anthropomorphism," and are reminded of Voltaire's saying: "*Si Dieu a fait l'homme en son image, l'homme lui a joliment rendu.*" But since we are mere human beings, it is absolutely impossible for us to think at all, save in human terms. There is, however, no danger of being misled. There is no danger of our regarding the First Cause as actually a man, or as less than a man, and all we have to do is, while using human language, to recognize its necessary inadequacy, and that it can but serve to indicate a true analogy. Evidently it is inexpressibly more true to speak of such First Cause as powerful than impotent, wise than foolish, and good than bad. But so to recognize the Great Cause of all things is enough for our practical purposes.

Before, however, proceeding to the practical deduction we desire to draw from this branch of "the helpful science," it may be well to turn back and reconsider the real nature of those self-evident primary truths the existence of which we have already recognized.

These truths, as we have seen, are self-evident and need no proof; while, at the same time, they are themselves indispensable in order that anything else may be proved. They are clear and certain, but they do not explain themselves. The rays of light they bring to the intellect are most luminous, but nevertheless they are rays, and their source—the luminary whence they radiate—remains hidden from our direct mental gaze, and can only be indirectly known through mediation and reflection. Truth, as we have seen, is a correspondence of thought with things. What, then, is and must be that with which these— at once the highest yet most fundamental of all truths—correspond? Upon them all the most certain deductions and inductions of science and all the most practical rules of art ultimately repose. The grade of perfection of every art and science depends directly upon and varies with the degree of accuracy wherewith our thoughts correspond with objective reality. But with what objective reality can those truths correspond which are absolute, universal, and necessary, except an absolute and infinite First Cause? They must reflect upon our minds illuminating rays from an inconceivable source of all light and of all truth — from God!

I am thus writing (as I said I should) not in the interest of any special form of religion, but merely as an earnest student of science and an uncompromising upholder of the dignity and worth of the human intellect. It is in the name of science, of reason, and in those names *alone*, that I have made, and again make, the declaration that the system of sensism and the folly of agnosticism are as devoid of a rational basis as they are morally pernicious; as also that the most exhaustive study of science—above all, of the science of sciences—gives us the most solid ground for affirming the existence of a most wise, powerful, and good God as the First Cause of the universe which surrounds us, and, therefore, of our own being. But our knowledge of Him is due to inference, not to intuition, and still less to any superrational, spiritual illumination. Of such illumination I, at least, possess no vestige, and my appeal as to Theism is only to the hard reason, the cold, dry intellect of my readers, and not at all to pious feelings or edifying sentiments.

Our moral intuitions thus proclaim the existence of God as the source and cause of such intuitions, but their absolute character shows us that they cannot depend upon even a Divine will, but must be of the very essence of that inconceivable Divinity itself. But although morality is absolute, and does not depend on God's will or our recognition of His existence, nevertheless that recognition most powerfully aids in and reinforces our ethical judgments. A climbing plant, though rooted in the soil, can live independently of bodies in its vicinity, yet it can never attain its perfection without external aid. So moral aspirations need for their full fruition the support of Theism. The same moral perceptions also confirm that dictate of our common-sense which declares that we have power over and are responsible for some of our actions. The very existence of a moral law implies and supposes the existence of some power on our part of obeying or disobeying it—that we have more choice as to at least some of the resolutions we form than a piece of paper thrown on the fire has a choice as to whether it will burn or not. For our reason clearly affirms that no one can be justly blamed or punished for doing anything he cannot possibly help doing. We may, of course, shoot a madman if we have no other means of saving our lives; but though we kill him, we do not think him wicked, but simply mad, and we may truly pity him all the time. But we well know that there are deeds which are really blameworthy, and there are few of us entirely free from self-reproach as to some actions of that kind.

When we recognize the certainty of God's existence in connection with our moral perceptions, the argument in favor of the existence of a future life becomes greatly strengthened. We said before that justice demands a future life; and our recognition of the being of God removes all difficulty with respect to its possibility.

That future existence can enable justice to be fully satisfied; and not justice only, but also the deepest and most vivid aspirations of our better nature.

But who can fulfil that expectation and aspiration of the enlightened conscience, or discern our true merits or demerits, but an infallible judge of conscience—one who has boundless wisdom, irresistible power, and is absolutely just? It must be He who can alone constitute our supreme good, and alienation from whom must therefore be our direct ill. It must be a First Cause who, being the Author of human nature, alone absolutely understands it. It must be God, the supreme Legislator, Judge, Rewarder, and Chastiser.

The truth that God exists is thus evident to whoever really understands what the idea of "duty" implies, and it will be the more evident to him the more fully that idea is understood.

There are three fundamentally different

kinds of greatness in the world: (1) *Material* greatness, which our senses can appreciate and imagine; (2) *Scientific* greatness, which can be apprehended by our intellect, but not by our sensitive faculty; and (3) *Moral* greatness, which can be adequately perceived by no human being, as we are not able accurately to perceive even our own merits and demerits.

As Pascal has well said, all merely material bodies—the earth and all the stars of the firmament—do not equal the value of one human mind; for it can know both them and itself, while such merely material bodies know nothing. It is no less clear that intelligence and knowledge, however great, are nothing when compared with goodness, and moral truth is out of all proportion the most important of all truth. Our reflective reason shows us that the true end of man is not that he should be strong, handsome, or learned, but that he should be "good." If a man were to obtain all possible knowledge, and possess an inventive genius placing the powers of nature at his disposal, yet, if he were not "good"—if he lived a life divorced from duty—he would have missed his true "end," and failed to "follow the right order." Ethical (moral) truths are, therefore, those which most concern us all, and are of all the most practical. The knowledge which helps us to be good is and must be more valuable than all other kinds of knowledge put together. This is the kind of "knowing" which helps the "doing" of that work which is set before every one of us to do.

The foregoing considerations show us how certain it is that no possible combination of circumstances can deprive us of our true end in life, and nothing but a bad will can divert us from it. However humble and obscure a man's lot in life may be, so long as he continues obedient to duty, it possesses a dignity and a worth which cannot be overestimated. Life, for him, is never without an adequate aim—never without utility, or deprived of a sphere of meritorious action. Further, no event in our lives can be without significance, and no action can be an indifferent one. A degree either of merit or of demerit, however minute and trifling it may be, must accompany every one of our conscious voluntary acts.

Such are the final considerations it is here desired to bring forward as being the most practical results of such study of "the helpful science." That science supplies us with a firm and sure support for all those motives which are most conducive to the welfare of the individual, the family, the State, and, ultimately, of all mankind.

Basing our edifice of rational thought upon the foundations afforded by the deepest and most certain truths accessible to the mind of man—truths upon which all science and all art must repose—we are enabled to recognize the beauty and majesty of human reason, whose main glory is the recognition of moral worth. The validity of our perception that there is a fundamental distinction between right and wrong and the authority of conscience are, as before said, guaranteed by the self-evidence of those fundamental verities which were recognized in the second and third parts of this essay. They are truths which can be recognized, and are fully within the grasp of every ordinary normal intelligence, although the subtle puzzles of skilful men may for a time blind many (as they have blinded and do blind many) to their certain truth. But since an accurate knowledge of such truths has the supreme importance here set forth, the responsibility of those who, by their teaching, would blind men to them, must be great indeed. Not less great is the obligation which binds men who see through such sophistic fallacies to stand up boldly and refute them, however great may be the hostility they may thereby bring upon themselves.

The system of *sensism*—that is, the system of the relativity of knowledge—possesses the gravest defects. It fails to attain the ultimate foundations of all our knowledge, or to account for and harmonize what consciousness tells us, by recognizing that substantial, persisting existence of which each man's consciousness, properly interrogated, cannot fail to assure him. It fails, also, to recognize the principle of contradiction, without which the intellect is reduced to a state of chaos. It is absolutely fatal to every germ of morality. It entirely negatives the first principles of all religion, and, finally, it stultifies itself by proclaiming its own untruth—a proclamation which is contained in its assertion that all our knowledge is but phenomenal and relative.

Intellectualism, on the other hand, lays its foundations in the deepest truths accessible to the human intellect, and accounts for and harmonizes the declarations of consciousness by recognizing our substantial and persistent being. It fully accepts the principle of contradiction, and thereby induces order into our thoughts and perceptions. It supports and enforces moral teaching, and firmly establishes the basis of all religion. Finally, it affirms and justifies its own truth in distinctly apprehending the declaration of our primary intuitions which contain their own self-evidence, and are recognized as so containing it.

How, then, can we account for sensism having attained the wide acceptance it has obtained, and how is it that men of undoubted intellectual distinction support it? We have already said that there are two good reasons which account for this. One reason is the course which human history has taken during those alternations of fashion, in intellectual as in other matters, which were called attention to in the first part of this essay.

As before pointed out, Descartes first broke with that philosophical system which had been elaborated by the keenest intellects of the Middle Ages, and sapped the basis of all certainty. Abandoning the conviction previously entertained that we have a direct perception of external objects, and basing his system exclusively on ideas, he paved the way for the whole range of modern philosophy. But mental images are but the means, not the objects of perception. We do not perceive "impressions," "images," or "representations" of the objects about us, but by means of them perceive the things themselves. An examination of what our own mind declares to us suffices to show that our faculties not only furnish us with images or impressions of things, but by means of those images and impressions, they *represent*—that is, they *make the thing present* to the intellect. When we visit a menagerie we do not perceive the sensations the animals within it produce on our organs of sense, but the animals themselves. Our sensations make bodies known to us without being themselves cognized. They, as it were, hide themselves from our notice in giving rise to the perception they elicit, and can only be detected by our expressly directing our attention to—turning our intellect upon—them. Through Descartes, and through the ambiguous system of Locke, came the long procession consisting of Spinoza, Berkeley, Hume, Kant, Fichte, Hegel, and Schelling, to Schopenhauer and Hartmann, with visions of rare and elaborate aerial palaces of cloudland, together with the less dazzling constructions of the brains of Mill, Bain, Spencer, etc., in our own day.

Thus our speculative contemporaries and our immediate predecessors may claim some indulgence for having accepted — taken for granted—a mode of thought so widely current, almost universally accepted, since Descartes's day. Nevertheless, it is the business of the philosopher to take no system for granted, and to bow to no authority. It was only by rebelling — after many doubts and much hesitation—against that of Descartes, and severely scrutinizing his position and arguments, that I at last obtained intellectual freedom and full mental satisfaction.

Thus it was that I became convinced the time had at last arrived for another renascence—for the rejection of all forms of Cartesianism, and the construction of a more satisfactory system based on principles the truth of which had been formerly recognized, but without any return to antiquated forms or obsolete prejudices. This is the system of intellectualism here put forward. It is a system which, unlike Cartesianism, rests not on ideas only, but also on those direct perceptions of objective fact which take us out of ourselves into that objective world which surrounds us on all sides, as every healthily constituted mind well knows.

History, then, largely helps us to account for the pernicious errors of those who advocate *sensism* and the relativity of knowledge. They are the victims of long-established prejudices which they have accepted unquestioningly.

There is another consideration, however, which concerns a temptation likely to be felt by some persons exceptionally gifted.

To excel in anything which is extraordinarily difficult must afford a special gratification to that love of superiority which so many men possess who prize the admiration of their fellows. And what could well be more gratifying to such a man than to have it believed that he apprehends things to which the common herd are blind, while he sees clearly that those things the vulgar regard as most certain (*e.g.*, their own continuous existence, etc.) are but so many delusions which his superior vision enables him to recognize as such and rise superior to. It is but too likely that this temptation has had its full effect on some of those who have startled the world by their bizarre philosophical notions.

There is yet another reason which it is to be feared has actuated, consciously or unconsciously, some members of the sensist school. A profound repugnance to religion, however discreetly or astutely veiled, is at the bottom of much of the popular metaphysical teaching now in vogue. *Delenda est Carthago!* No system is to be tolerated which proclaims the existence of a personal God, moral responsibility, and future rewards and punishments.

Men who feel thus, and who deal with metaphysics, become driven from one absurdity to another in their struggle to maintain an essentially irrational negative position.

The fundamental principle of all religion, the existence of God, reposes, as we have seen, on a series of self-evident truths, which must one after another be denied if the existence of God is to be impugned. Accordingly, because morality is the most important support to Theism, any real distinction between utility and virtue must be denied. Also, since the principle of causation establishes the Theistic position, it is necessary to deny the validity of that principle.

But the principle of causation is a self-evident and necessary truth; therefore it becomes incumbent on those who deny the certainty of our knowledge of God's existence also to deny that we can know any self-evident and necessary truths at all—therefore the principle of contradiction has to be denied along with that of causation.

But those who deny this are logically compelled to go further still, and deny the possibility of our knowing objective truth at all. So to deny, however, involves the denial of the validity of memory, which (as we have seen) introduces us to a knowledge of objectivity, and this denial carries with it that of any knowledge of our own existence.

Even thus the spirit of irrational negation does not reach its climax. Not only is it necessary for these men (if they would attain their end) to affirm the certainty of uncertainty; the necessity of the truth which declares there is no necessary truth; to make use of their memory in order to deny the validity of memory, and to know their continuous existence long enough to be able to deny the possibility of their knowledge of their having any knowledge of it; but they must go yet further, and deny the real existence of even thought itself. At a meeting of a metaphysical society in London, it was not long ago, "thought" was expressly declared to be a "misleading term, the use of which should be carefully avoided"—and this in spite of the fact that it is utterly impossible to form any kind of judgment except by the aid of "thought," and by practically admitting its unique value. What are we to think as to the merits of men who promulgate doctrines the consequences of which are so extremely anti-social? Can we say they are only foolish?

One thing at any rate is certain. They are unable to reply to such criticisms and objections as are here urged. It is some years since a refutation of the system of sensism and the relativity of knowledge was published by the present writer. Yet not one of his critics has ever ventured to tackle the arguments therein put forward, or to try and defend his untenable position.

But "the helpful science" does not only serve as a practical guide in life by showing us the true end of human existence, but also shows us that the whole world is replete with purpose, as Emerson has sung.

Since the Universe proclaims to human reason that it has for its First Cause a Being possessing power, will, intelligence, and goodness (such as are inadequately shadowed forth in our own faculties), it follows that such a First Cause must be the author of yet another kind of causation, namely, *final causation*. Philosophy assures us that there is a reason *why* things are, and why they are such as they are. Can the science of sciences also help us to an answer to this riddle?

Directing our gaze on the world of animals and plants—or on organic nature—we find that each organism plainly declares that it is the embodiment of both an internal and an external "purpose." Every organism, in that it *is* an organism, is a structure the parts of which are reciprocally "ends" and "means" directed to the conservation of the individual. But each organism also proclaims itself to be instinct with a finality not its own—a purpose beyond itself—which penetrates and regulates the inmost recesses of its structure.

The phenomena which attend the development of the adult structure from out of its embryonic germ plainly announce that they are directed towards a future end. As Claude Bernard has said, in every living germ there is a directing idea which develops and manifests itself in and by that germ's gradual organization. This truth is especially manifested in those animal instincts by which, in such curious and complex ways, a parent insect provides for the future welfare of a progeny it is destined never to behold. But a careful study of nature also serves to reveal to us that it consists of a hierarchy of energies and faculties; so that a successively increasing fulfilment of "purpose" runs through the irrational creation up to man.

There can be little doubt but that before

life existed on this globe many ages passed by while it was but a mass of inorganic substances, each possessing its own powers and energies. Evidently, also, the earth was clothed with vegetation long before its groves were peopled with the higher forms of animal life. For thousands, perhaps millions, of years its groves were songless, and the flowers of its herbs were insignificant and inodorous. For all we know such life might have continued until now, for plants can thrive without the presence of animals. The animal world, however, could not exist without a world of plant-life, for from plants all animals must, directly or indirectly, acquire their nutriment.

We now know how wonderfully rich, in both animals and plants, the world was in what is called the Miocene period. Nor can we find any reason why such organisms should not continue forever to live on without the risk of subjugation by a rational creature, such as man.

When at last men appeared they constituted a race which, though capable of living on vegetable food alone, yet needed animals —dogs, horses, flocks and herds—to attain any high development and complex social organization.

Thus an increasing service, and consequently an increasing dependence, runs on from the world of lifeless seas and rivers, rocky mountains and sandy wastes, through vegetable and animal life up to man. The sun's beams, though potent in inducing atmospheric and aqueous currents, nevertheless actually serve the organic world far more than the inorganic, and the powers of nature do more for the animal, with all its sentient faculties, than for the plant. But they do most of all for man, since he makes use of all the lower orders of existence, both organic and inorganic. Therefore we must affirm that God has evidently willed most service to man of all His earthly creatures.

Whatever purposes these two great groups —animals and plants—may serve, they must be affirmed to exist specially for man, since it is a fact that he forms the culmination of all those different kinds of creatures which have been caused to exist on the surface of our planet, and, as a fact, derives most service from them.

As to the end of human life itself, we have already seen that it is and must be the fulfilment of duty—the exercise of will according to the dictates of right reason. Therefore, since the lower forms of life subserve his being, and since the world embodies the will of an Intelligent First Cause, we may, with certainty and without exaggeration, declare that when the first and lowest forms of life made their appearance in the world, their highest conceivable purpose and meaning was to bring about the fulfilment of the moral law—a fulfilment bearing fruit beyond the life of which we have experience.

So much certainly follows as a deduction from the facts and principles which "the helpful science" sets before us. Can mere speculation carry us a step further as to the wherefore of the world's existence? There is one conception which does seem to bring us even nearer to God than the conception "duty," and that is "love." Love is the noblest, most self-sacrificing, most tender, yet most strong, energetic, and unflinching of all human energies, and it is that which binds us to all that is most admirable and beautiful in aspect, character, and conduct.

May we not, then, believe that through it we reach the truest analogy, possible for us, with the ultimate creative purpose of the Divine First Cause—an analogy perceptible to us by means of those certain and evident intuitions through which God has deigned to illuminate the intellect of man? Such, if it cannot be said to be the positive teaching, would appear to be the most probable suggestion which arises from the various considerations which have been here successively brought forward, and which demonstrates the power and dignity of human nature.

And now, in concluding, I venture to express a hope that the utility and importance of "the helpful science" has been brought home to the minds of such of my readers as may not before have appreciated those qualities as belonging to it. It has, I hope, been clearly shown that the first principles of religion, morals, art, and science, all repose upon a common foundation which the "science of sciences" makes known to all those who are willing patiently to undertake the needful self-interrogations and observations of things external.

By the term "a common foundation," however, I am far indeed from meaning thereby that all our knowledge reposes upon any one principle; which was the error of Descartes and all his numerous followers. "The helpful science," as it aids us in different kinds of knowledge and different kinds of activity, so itself demands for its support the exercise of very different faculties.

It reposes on the exercise of our sensitivity, and our sense-perceptions; on our direct intellectual intuition and the declarations of reflective thought; on the observation of the bodies which make up the material world, and upon the information we derive from our fellow-men. It is emphatically a human science, and responds to the needs of an intellect such as ours, which, wonderful and highly endowed as it is, is none the less essentially limited, as is shown even by the need we so often have of going through processes of reasoning to arrive at truth. Being so limited it needs corresponding aid, and such aid has been duly supplied to it through the ministry of the "helpful science."

THE END

MENTAL AND MORAL SCIENCE

THE PRINCIPLES OF ETHICS.

By BORDEN P. BOWNE, Professor of Philosophy in Boston University.   8vo, Cloth, $1 75.

Shows uncommon clearness, penetration, and breadth of outlook, along with a certain practical good sense which keeps the writer from being tricked by mere words.—*Advance*, Chicago.

It is written with the clearness and incisiveness of the author's best style, and will clear the minds of many readers of confusion on serious points.—FRANKLIN CARTER, President of Williams College.

METAPHYSICS.

A Study in First Principles.   By BORDEN P. BOWNE. 8vo, Cloth, $1 75.

It will take high rank in the philosophical literature of the time.— *N. Y. Tribune.*

We are free to express the opinion that the work has real value. The conclusions which he reaches will be found especially useful to those who find themselves caught in the drift of materialism.—*N. Y. Herald.*

PHILOSOPHY OF THEISM.

By BORDEN P. BOWNE.   8vo, Cloth, $1 75.

One of the simplest in statement and clearest in thought of the many works on the subject.—*Critic*, N. Y.

Epigrammatic terseness and fine scorn are the delightful concomitants of lucid reasoning and clear arrangement.—*Evangelist*, N. Y.

Professor Bowne is an acute and original thinker and profound logician.—*Albany Press.*

INTRODUCTION TO PSYCHOLOGICAL THEORY.

By BORDEN P. BOWNE.   8vo, Cloth, $1 75.

This is not a dogmatic treatise of empirical psychology, much less a digest of physiological psychology and the fanciful theories that cluster round that shadowy border-land of research, but a series of essays in pure psychology, the basis of the whole performance being facts, not theories.—*Boston Beacon.*

PSYCHOLOGY.

By JOHN DEWEY, Ph.D. 12mo, Cloth, $1 25.

As a philosophical text-book its claims to the recognition of thinkers are very great, while as an exposition of one of the most interesting of sciences it will be a hand-book of inestimable value to students.—*Commonwealth*, Boston.

The book, I think, marks a distinct progress on anything, in that department, which has been done in English, in which most writers have really omitted the part of Hamlet in their psychological drama.—EDWARD CAIRD, Professor of Moral Philosophy in the University of Glasgow.

THE ELEMENTS OF DEDUCTIVE LOGIC.

By NOAH K. DAVIS, Ph.D., LL.D., Professor of Moral Philosophy in the University of Virginia. Post 8vo, Cloth, 90 cents.

The definitions are good, and the expositions of the difficulties of the subject are accomplished with enough fulness to be intelligible without confusion.—*Independent*, N. Y.

I am not acquainted with any treatise on logic that contains within the same compass so much sound logical doctrine so perspicuously expressed. It would not be difficult to point out in this small work at least half a dozen distinct gains to the science.—Professor COLLINS DENNY, Vanderbilt University, Nashville, Tenn.

The Elements of Inductive Logic, by the same Author, is in Press.

THE THEORY OF THOUGHT.

A Treatise on Deductive Logic. By NOAH K. DAVIS, Ph.D., LL.D. 8vo, Cloth, $2 00.

A comprehensive account of the science of Logic from its earliest days, with every variety of example to illustrate the principles. . . . The author is to be commended for his industry, his earnestness, his intelligence in the arrangement of his material, and the general excellence of his literary style.—*Philadelphia Evening Bulletin*.

Published by HARPER & BROTHERS, New York

☞ *The above books will be sent to any address in the United States, Canada, or Mexico, on receipt of price. If ordered sent by mail*, 10 *per cent. should be added for postage.*

www.ingramcontent.com/pod-product-compliance
Lightning Source LLC
Chambersburg PA
CBHW032152160426
43197CB00008B/876